Nuclear Test Ban

Nuclear Test Ban

Converting Political Visions to Reality

Ola Dahlman

Svein Mykkeltveit

and

Hein Haak

Foreword by Maxime Verhagen, Jonas Gahr Støre and Carl Bildt

 Springer

Ola Dahlman
Fredrikshovsgatan 8
SE-115 23 Stockholm
Sweden
ola.dahlman@gmail.com

Svein Mykkeltveit
NORSAR
POB 53
NO-2027 Kjeller
Norway
svein@norsar.no

Hein Haak
Royal Netherlands Meteorological
Institute (KNMI)
POB 201
3730 AE De Bilt
The Netherlands
hein.haak@knmi.nl

Cover illustrations: Nevada Test Site. The subsidence craters in this area are caused by nuclear testing. Photo courtesy of National Nuclear Security Administration/Nevada Site Office Atmospheric thermonuclear test explosion "Truckee". Photo from the Nuclear Weapon Archive.

ISBN 978-1-4020-6883-6   e-ISBN 978-1-4020-6885-0

DOI 10.1007/978-1-4020-6885-0

Library of Congress Control Number: 2008937915

Printed on acid-free paper

9 8 7 6 5 4 3 2 1

springer.com

Foreword

Nuclear tests have caused public concern ever since the first such test was conducted, more than six decades ago. During the Cold War, however, conditions were not conducive to discussing a complete ban on nuclear testing. It was not until 1993 that negotiations on such a treaty finally got under way. From then on, things moved relatively quickly: in 1996, the United Nations General Assembly adopted the Comprehensive Nuclear-Test-Ban Treaty (CTBT). To date, the Treaty has been signed by 178 states and ratified by 144, though it has yet to enter into force, as nine out of 44 "Annex 2 states", whose ratification is mandatory, have not heeded the call. Nevertheless, the CTBT verification system is already provisionally operational and has proven its effectiveness. We commend the CTBT organisation in Vienna for its successful efforts to build a verification network.

This book is an excellent overview of the evolution of the CTBT and its verification regime. The authors are eminent scholars from the Netherlands, Norway and Sweden who have been intimately involved with the CTBT and its verification agency, the CTBTO Preparatory Commission, from their inception to the present day. They have written a thorough and engaging narrative of the long road that led to the CTBT. Their story will appeal to both the layman and the expert and provide useful lessons for future negotiations on disarmament issues.

We believe that the call for disarmament and non-proliferation is gaining renewed momentum in the 21st century. The proliferation of weapons of mass destruction is one of the main threats facing the world today. It should be prevented through a legally binding system of international treaties and enforcement mechanisms. The Nuclear Non-Proliferation Treaty (NPT), which was signed in 1968, aims not only to prevent the proliferation of nuclear weapons but indeed to eliminate them entirely. The CTBT supports the NPT in this respect. Further proliferation would be detrimental to our common security interests and the world as a whole. We all need to intensify our efforts to eliminate this threat.

The Netherlands, Norway and Sweden support a world free of nuclear weapons, though we realise that this will not happen overnight. A legally binding nuclear test ban will be an obstacle to nuclear proliferation and the onset of a new nuclear arms race. We therefore call on the governments concerned to ratify this crucial treaty as an important step towards ridding the world of nuclear weapons and the threat these weapons pose to us all.

Maxime Verhagen
Minister of Foreign Affairs, The Netherlands

Jonas Gahr Støre
Minister of Foreign Affairs, Norway

Carl Bildt
Minister for Foreign Affairs, Sweden

May 2008

Preface

On July 16, 1945 the world changed and on August 6 the world knew. The "Trinity" nuclear test in the Alamogordo desert in New Mexico, USA and the destruction of the Japanese city of Hiroshima and that of Nagasaki three days later marked the beginning of the nuclear weapons era, an era when nuclear weapons dominated relations between the main powers and international security policy in general. More than 2000 nuclear test explosions have been conducted by eight States to support the development of a large number of nuclear weapon systems.

The fall of the Soviet Union and the 9/11 terror event in 2001 changed the focus, and the role of nuclear weapons in the security relations among States has diminished. Nevertheless, large numbers of nuclear weapons are still maintained, some on hair-trigger alert, and the issue of non-proliferation is high on the international agenda. The Non-Proliferation Treaty (NPT) and the Comprehensive Nuclear-Test-Ban Treaty (CTBT), together with a number of regional nuclear-free zone agreements, are key elements of the non-proliferation regime.

The signing of the CTBT, fifty years after the first initiatives to limit nuclear testing were taken, was an important political achievement. Not only is the CTBT an essential element in a global non-proliferation regime and a step towards nuclear disarmament, the conclusion of the Treaty also removed a longstanding hurdle en route to further steps towards nuclear arms control and disarmament.

The signing of the CTBT marked the end of a very long journey, but it also proved to be the beginning of another. The Treaty was opened for signature in 1996, but it is still not in force. Politically it was put on the back burner in 1999, when the USA failed to ratify it. Eight more States – China, Egypt, India, Indonesia, Iran, Israel, North Korea and Pakistan – also need to ratify before the Treaty enters into force.

Despite the political difficulties, implementation of the verification regime, the most elaborate ever created, is approaching its final stage. The aim of these verification measures is to give all States an equal and fair possibility to monitor compliance with the Treaty by providing high quality information to all States. The future CTBT Organization (CTBTO) does not, however, have the authority to draw conclusions regarding the nature of the events that are observed. This political step rests with States Parties to the Treaty. To make the Treaty and its verification regime a truly global undertaking it is thus necessary to increase engagement to ascertain that the competence and resources needed are available to States Parties when the Treaty enters into force.

More than 300 monitoring stations in 89 countries around the world, together with an intrusive on-site inspection regime, are being implemented. Together, these not only create a high performance verification system, but also demonstrate that extensive verification arrangements can be established. This regime has now been under implementation over a period of more than ten years and, provided States allocate the necessary funds and that the CTBTO Preparatory Commission focuses its work on the remaining key elements of the system, we should by now have passed the time when there is any uncertainty about the system's readiness when the Treaty enters into force.

During its establishment the monitoring system has benefited from more than ten years of rapid development in science and technology. The dramatic development of information technology offers great potential to further upgrade the data analysis procedures and to improve the quality of the products provided to States. The verification system was regarded as adequate when the Treaty was signed. Developments since then have enhanced confidence in the verification system and its ability to provide data to allow States to adequately verify compliance with the CTBT.

The rapid developments of our societies over the last century are based on unprecedented achievements in science and technology, which have shaped our modern societies and have also been extensively used to enhance the capability of the military component of our security system. How can science and technology play an equally strong role in promoting security in this new, broad perspective? CTBT is a good example of the successful application of science and technology to global, non-military security issues: how do we carry this good example further into new domains?

As in the case of CTBT, verification arrangements might also in the future be both extensive and complex and would thus not be an element that could be added at the end of a negotiation. On the contrary, a credible verification regime might well be a prerequisite for an agreement. The scientific and technical preparatory work carried out by the Conference on Disarmament's (CD) Group of Scientific Experts during 1976–1996 greatly facilitated the negotiation and the implementation of the CTBT. Such international scientific and technical cooperation could also help create mutual confidence. In our view we should aim at creating international mechanisms by which technical and other non-political issues, related to our security agenda, could be explored at an early stage. The Intergovernmental Panel on Climate Change, IPCC, which shared the 2007 Nobel Peace Prize, is a good example of such a long-term scientific effort, on a global scale, to address a most important security issue.

The threat to our environment is an example showing that the security perspective of today and tomorrow is broader than that of yesterday. Until a few years ago, a State's security was synonymous with a strong military defence, and ever since they were created States have invested heavily in the military component of their security regimes. The security concept is different today and requires new strategies, national and international, for investing in security.

This book is about investments in nuclear disarmament and non-proliferation through over half a century of efforts to ban the testing of nuclear weapons, the most deadly weapons ever invented. We feel that the time is becoming ripe to bring the Treaty back on track and we see an urgent need to inject new energy into the process of bringing the CTBT into force. This book is a modest contribution to this process.

We approach the CTBT and its implementation from three different perspectives; political, scientific and managerial. Our ambition is not only to describe what has happened and what we have witnessed over almost forty years of experience with the test ban issue, but also to comment and reflect and to identify lessons learned. Lessons that might prove useful not only in bringing the CTBT into force, but also when negotiating and implementing future disarmament and other security-related treaties. Our ambition is to provide a fair, unbiased description of what we have experienced. The comments and reflections are personal, colored by our profound engagement with the CTBT and by our belief that multilateral treaties are an important element in building global security.

This book is intended for professionals in the political, diplomatic, scientific and military areas, dealing with arms control and disarmament. It is also intended for NGOs and journalists seeking a deeper understanding of the nuclear test ban issue and of multilateral arms control and disarmament treaties in general. The book could serve as an introduction to diplomats and experts about to start working on the implementation of the CTBT or on the negotiation or implementation of other security-related, multilateral treaties. It could also be used as a textbook for training young diplomats and other experts in arms control and disarmament.

The book can be divided into four parts. Two, quite different initial background chapters first describe nuclear testing, nuclear weapons and earlier treaties in the nuclear area, followed by the seismological, hydroacoustic, infrasound and radionuclide monitoring technologies, which in the CTBT verification regime are integrated to form the most comprehensive global verification system ever created. Chapters 3 and 4 summarize and discuss the CTBT negotiations and the Treaty itself. The main part of the book, Chapters 5, 6, 7, 8, 9, 10, covers the implementation of the Treaty from different perspectives. A description of the work during ten years within the framework of the CTBTO Preparatory Commission is given in Chapters 5 and 6. Chapter 7 reviews important results on the capabilities observed during the testing of the system. Chapter 8 discusses efforts at a national level to support implementation. Chapter 9 gives our analysis of the CTBTO Preparatory Commission and its Provisional Technical Secretariat from an organizational and managerial perspective. In Chapter 10 we discuss the CTBTO Preparatory Commission in relation to those international actors, in particular the States Signatories, with which it interacts. Finally, in Chapter 11 we provide our analysis of the experiences gained and of the lessons learned. Our assessments are made from political, scientific and managerial perspectives. Chapter 11 also includes some reflections on how to bring the CTBT back on track. As the CTBT is an excellent example of a non-military application of science and technology in support of global security, we finally briefly discuss the new security agenda and the need for a new strategy to invest in security.

The chapters are written so that later chapters build on explanations given in earlier ones. The link, however, is not too strong and it should be possible to read individual chapters of particular interest. The last chapter is self-contained and can be read independently of the others.

We have been closely engaged with the test ban issue for several decades. We were, for a long period of time, deeply involved in CTBT negotiations at the CD in Geneva and in its Group of Scientific Experts, tasked with developing a test ban verification system. During the implementation of the Treaty we have all been deeply involved with the CTBTO Preparatory Commission since its establishment in 1996.

During all these years we have been privileged to work with a large number of scientific colleagues and diplomats in Geneva and Vienna and at a number of research institutions around the world. They have all, in one way or another, influenced our thinking and contributed to this work. In particular we would like to mention our close friends in the Group of Scientific Experts at the CD in Geneva and in Working Group B on verification in the CTBTO Preparatory Commission in Vienna. We would also like to thank our friends and colleagues at the Provisional Technical Secretariat (PTS) of the CTBTO Preparatory Commission. Over the years we have enjoyed most stimulating interactions with Wolfgang Hoffmann, who served as the Executive Secretary of the CTBTO Preparatory Commission for most of the period covered by this book; and with his successor Tibor Tóth. Wolfgang Hoffmann has also provided valuable comments on the book. We would also like to thank Tibor Tóth and the PTS staff for generously providing illustrations and specific information on the PTS.

A number of colleagues and friends have made most valuable contributions to the book. Lars-Erik De Geer, an outstanding expert on radionuclide monitoring, has helped us out with those parts of the book. He has also contributed a large number of most valuable suggestions and comments on other parts. We deeply appreciate Lars-Erik's support, which goes far beyond what could be expected, even from a good friend.

Jenifer Mackby and Peter Basham have reviewed the full manuscript. Their corrections and comments have certainly improved the text. Bernard Massinon and Victoria Oancea have reviewed large parts of the text and provided a number of helpful comments. Frode Ringdal, who served as the scientific secretary to the Group of Scientific Experts, has provided valuable suggestions on how to reflect the work of that group. We also appreciate valuable suggestions from Jay Zucca.

Many of our PTS friends have helped us out on a number of issues. Annika Thunborg, head of the PTS Public Information Section, has been instrumental in providing illustrations and photos used throughout the book. She has also provided most valuable comments on many parts of the book, based on her prior diplomatic experience. John Sequeira, Director of the Administrative Division, has provided material on personnel and financial matters and helped us to put this and some rules and regulations into the right perspective. He also provided a number of valuable comments on our text. Jerry Carter has commented on parts of the text, coordinated various contributions from the IDC Division, and provided information on services to States Signatories. Aili Bi, with the External Relation Section and with a background in the CTBT negotiations, has also suggested a number of improvements to the overall structure of the book. Mordechai Melamud has helped with important comments on aspects of the PTS work related to the on-site inspections regime. Other PTS employees have contributed valuable assistance in the form of information on specific issues, as well as support with illustration material. In this regard we appreciate contributions from Michael Akrawy, Arne Bell, Paola Campus, Luis Cella, John Coyne, Peter Hulsroj, Don Phillips, Stefka Stefanova, Todd Vincent and Lassina Zerbo.

We thank colleagues at our home institutes for their support and encouragement. Steven Gibbons, Tormod Kværna, Frode Ringdal and Johannes Schweitzer, all at NORSAR, are thanked for text reviews, stimulating discussions and help with illustrative material. At KNMI Läslo Evers, Bernard Dost and Ko van Gend supplied figures, and Marieke Laagland is thanked for reformatting and correcting all the text. We thank KNMI for financial support to cover printing charges for an all color book. We thank Ben Maathuis for providing illustrative material.

The responsibility for whatever errors or ambiguities that may remain rests solely with us. The views presented in this book are those of the authors and may not reflect, and do not represent, those of the institutions, organizations and national authorities with which the authors are or have been associated.

Stockholm, Sweden   Ola Dahlman
Kjeller, Norway   Svein Mykkeltveit
De Bilt, The Netherlands   Hein Haak

Contents

xvi   Contents

Chapter 1
To Test or Not to Test...

On July 16, 1945 the world changed and on August 6 the world knew. The "Trinity" nuclear test in the Alamogordo desert in New Mexico, USA marked the beginning of the military nuclear era. The destruction of the Japanese cities of Hiroshima on August 6 and of Nagasaki on August 9, 1945 demonstrated the overwhelming power of this new weapon. These explosions marked the end of the Second World War and the beginning of a nuclear arms race: an arms race that divided the world between those having nuclear weapons and those that decided not to have them or that so far have been unable to acquire them. Eight States have conducted in all more than 2000 nuclear test explosions involving some 2400 nuclear devices. The USA and the former Soviet Union have carried out the lion's share of these tests. Over the years, nuclear testing has been an essential element in the development of nuclear weapons and weapon systems. Numerous such systems have been developed and deployed and, despite reductions during the last two decades, large numbers of nuclear weapons are still deployed, some on hair-trigger alert, or stockpiled. The US Arms Control Association estimates (ACA 2007) that the total number of nuclear warheads, operational or stockpiled, on our globe today exceeds 30 000, the vast majority being in the hands of Russia and the USA.

Since the very first nuclear explosion, nuclear weapons have dominated relations between the main powers and international security policy in general. These weapons also led to the development of new military doctrines. The one on Mutual Assured Destruction, with the fitting acronym MAD (Sokolski 2004), is the most controversial. The fall of the Soviet Union and the terror event of 9/11 2001 changed the focus of security politics, and the role of nuclear weapons in the security relations among States has diminished. The issues of nuclear disarmament and non-proliferation are, however, important element in today's international politics.

The nuclear arms race also initiated efforts to limit the spread and development of such weapons with the ultimate goal of eliminating them. A large number of initiatives were taken to limit or ban nuclear testing before the Comprehensive Nuclear-Test-Ban Treaty (CTBT) was finally signed in 1996. The Partial Test Ban Treaty (PTBT) (Kimball and Boese 2003), banning nuclear explosions in space, in the atmosphere and under water, was signed by

O. Dahlman et al., *Nuclear Test Ban*, DOI 10.1007/978-1-4020-6885-0_1,
© Springer Science+Business Media B.V. 2009

the UK, USA and USSR in August 1963. In 1974 the USA and USSR concluded the bilateral Threshold Test Ban Treaty (TTBT) (FAS 1990), limiting nuclear weapons tests to yields below the equivalent of 150 000 tons (150 kt) of the explosive TNT. The Non-Proliferation Treaty (NPT) (IAEA 1970) of 1968, which aimed at preventing the spread of nuclear weapons and at promoting nuclear disarmament, is considered the cornerstone of the nuclear non-proliferation regime. A number of regional nuclear-free zone treaties have also been concluded to strengthen non-proliferation of nuclear weapons. The final goal, to put the genie back into the bottle by halting the nuclear arms race and achieving nuclear disarmament, is still very remote.

1.1 Testing History – More than 2000 Nuclear Explosions

The Non-Proliferation Treaty identifies five Nuclear Weapons States: China, France, United Kingdom, USA and USSR. Since the NPT entered into force in 1970 three more States have acknowledged that they have conducted nuclear test explosions: India, North Korea and Pakistan. South Africa has announced that they have developed and built, but did not test, six nuclear weapons, which they later destroyed, and they have now renounced nuclear weapons (Stumpf 1995). It is generally assumed that Israel also possesses nuclear weapons but no test explosion has been attributed to Israel. It has been the policy of the Israeli authorities to maintain ambiguity about the possession of nuclear weapons.

Statistics on nuclear testing are available from several sources, giving marginally different numbers. The statistics presented in Table 1.1 are taken from the references given in the table. The table gives the reported number of devices detonated and the number of tests conducted. A number of the tests are

Table 1.1 Summary of nuclear testing (as of February 2008)

State	First test	First thermonuclear test	Number of tests	Number of devices
USA[1]	July 16, 1945	1952	1030	1125
Soviet Union[2]	August 29, 1949	1955	715	969
United Kingdom[3]	October 3, 1952	1957	45	45
France[3]	February 13, 1960	1968	210	210
China[3]	October 16, 1964	1967	45	45
India[4]	May 18, 1974		3	6
Pakistan[4]	May 28, 1998		2	6
North Korea[5]	October 9, 2006		1	1
Total			2051	2407

[1] (DOE 2000), [2] (Mikhailov 1996), [3] (NRDC Archive 2007), [4] (NRDC 2007), [5] (NRDC 2006)

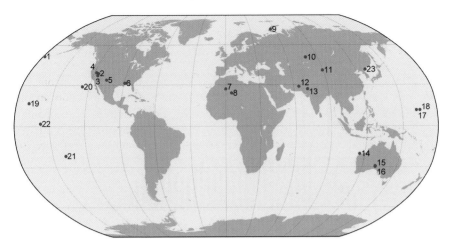

Fig. 1.1 Test sites of the world. 1. Amchitka Island, Alaska (USA), 2. Central Nevada Test Area, Nevada (USA), 3. Nevada Test Site, Nevada (USA), 4. Fallon, Nevada (USA), 5. Trinity Site, New Mexico (USA), 6. Hattiesburg, Mississippi (USA), 7. Reggan, Sahara Desert, Algeria (France), 8. In Ekker, Algeria (France), 9. Novaya Zemlya (USSR), 10. Semipalatinsk (USSR), 11. Lop Nor, Western China (China), 12. Chagai Hills, Baluchistan (Pakistan), 13. Pokharan, Rajastan Desert (India), 14. Monte Bello Islands, Australia (UK), 15. Emu Field, Australia (UK), 16. Maralinga and Woomera Test Sites, Australia (UK), 17. Eniwetok Atoll, Marshall Islands (USA), 18. Bikini Atoll, Marshall Islands (USA), 19. Johnston Island (USA), 20. Eastern Pacific Ocean (USA), 21. CEP, Mururoa and Fangataufa Atolls, French Polynesia (France), 22. Christmas Island, Kiribati (UK and USA), 23. P'unggye (North Korea). In addition, this list can be supplemented with other locations in the world where nuclear explosions have occurred, such as Hiroshima and Nagasaki in Japan under wartime conditions, and some locations in the South Atlantic Ocean, where the USA carried out three atmospheric tests (Operation Argus) in 1958. Peaceful Nuclear Explosions (PNEs) in the USA were carried out at Farmington, New Mexico, at Carlsbad, New Mexico, at Green Valley, Colorado, and at Rifle, Colorado. Sites of PNEs carried out by the former Soviet Union are shown in Fig. 1.3

reported to involve two or more nuclear devices detonated simultaneously. According to the sources used, in all 2407 nuclear devices have been exploded in 2051 nuclear test explosions. Of these, 551 took place in the atmosphere or under water and 1500 underground. Figure 1.1 shows the nuclear weapons test sites. Figure 1.2 shows the number of tests over the years for the States that have carried out tests.

1.1.1 China

China conducted its first nuclear test explosion on October 16, 1964 in the atmosphere close to the Lop Nor lake in the Xinjiang Autonomous Region. This area has been the testing ground for all its 45 nuclear explosions, with a

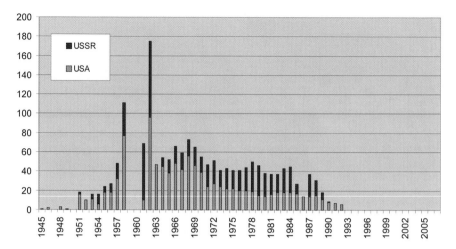

Fig. 1.2a Number of nuclear tests performed by the USA and USSR, 1945–2007

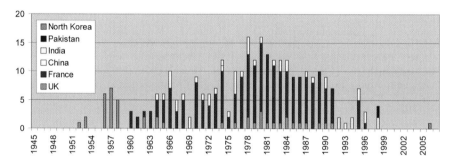

Fig. 1.2b Number of nuclear tests performed by the United Kingdom, France, China, India, Pakistan and North Korea, 1945–2007

testing schedule of one to two explosions per year. China has conducted 22 explosions in the atmosphere, the last one in October of 1980. 23 explosions were carried out underground, the first one in September 1969. China conducted its first test of a thermonuclear weapon in 1967. It is interesting to note that the last Chinese explosion took place as late as 29 July 1996, less than a month and a half before the CTBT was adopted at the UN General Assembly.

1.1.2 France

Starting on February 13, 1960, France has conducted in all 210 nuclear explosions, 50 in the atmosphere and 160 underground. The initial four tests in 1960 and 1961 were conducted in the atmosphere at a test site in Algeria, to be followed by 13 underground explosions, also conducted in Algeria. In 1966 France moved her testing to the Mururoa and Fangataufa Atolls in the Tuamotu Archipelago in the Pacific, where 46 atmospheric explosions were conducted until 1974. The first thermonuclear test was conducted in 1968. In 1975 France moved testing underground and carried out another 147 tests before testing ended on January 27, 1996. It should be noted that France also conducted some last-minute tests. After a break in testing from July 1991 until September 1995, France carried out a series of six tests during the end of 1995 and the beginning of 1996. After the last test France dismantled her Pacific testing facility.

1.1.3 India

India conducted its first nuclear explosion on May 18, 1974 at the Pokhran test site in Rajasthan province in the north-western part of India. The event was described by the Indian government as a peaceful nuclear explosion. 24 years later, on May 11, 1998, India resumed its testing at the same site and announced that three nuclear devices were detonated simultaneously underground. India claims that one of the devices was thermonuclear, a claim questioned by some experts (India Nuclear Update 2003). Two days later, on May 13, 1998, India announced that it had simultaneously (Operation Shakti 1998) detonated two additional low-yield devices. After this test series India announced a moratorium on nuclear testing.

1.1.4 North Korea

North Korea announced the conduct of an underground nuclear explosion on October 9, 2006. The yield of the test was estimated from seismic observations to be less than 1 kt. This is an order of magnitude lower than might be expected from a first nuclear test (see Box 1 on nuclear devices).

1.1.5 Pakistan

Pakistan's authorities announced that five nuclear devices were detonated simultaneously in a test conducted on May 28, 1998. An additional test was carried out on May 30, 1998. The tests took place in the province of Baluchistan in south-western Pakistan. Seismic observations confirm that an explosion occurred on May 28. These observations cannot identify the number of devices detonated, nor can they confirm that the explosion was nuclear.

Some doubt has been expressed about the number of devices detonated (Bidwai and Vanaik 2000). Pakistan too announced a testing moratorium after these tests.

1.1.6 Soviet Union

The Soviet Union carried out its first test on August 29, 1949. 715 tests, involving 969 nuclear devices in all, have been reported: 214 in the atmosphere, five under water and 496 underground. With the exception of a few early atmospheric explosions, the Soviet Union has used two areas for her nuclear weapon tests, one close to Semipalatinsk in what is today Kazakhstan and one at the Novaya Zemlya Islands in the Arctic Ocean. The last test was carried out in October 1990, before the disintegration of the Soviet Union. Atmospheric testing ended in 1962 and the first underground test was conducted in 1957. The USA and the Soviet Union had a test moratorium from the beginning of 1959 until the autumn of 1961. Following the end of that moratorium, the Soviet Union and the USA embarked on intense testing programs. In the autumn of 1961 the Soviet Union carried out 59 test explosions during a period of little more than three months, including the huge explosion of October 30, 1961, the most powerful explosion ever carried out, with an estimated yield of close to 60 million tons (60 Mt), or about 4000 times the yield of the bomb used on Hiroshima. The intense testing activity continued in 1962 with an additional 79 tests.

The Soviet Union had an extensive program to explore the possible use of explosions for civilian purposes, usually referred to as Peaceful Nuclear Explosions (PNEs) (Nordyke 2000). 124 such explosions have been reported (Mikhailov 1996), some involving multiple devices, conducted at a large number of sites across the Soviet Union during the years 1965–1988. Figure 1.3 shows the locations of the PNE explosion sites in the former Soviet Union. The explosions were conducted for a variety of purposes: to make underground storage space, to put out fires in gas fields, to stimulate oil and gas recovery, to create dams, and as sources for long-range seismic exploration profiles.

1.1.7 United Kingdom

The United Kingdom has been reported to have conducted 45 explosions, 21 in the atmosphere and 24 underground. The first UK test was conducted on October 3, 1952 at Monte Bello Island, Australia. The UK conducted in all 12 explosions in the atmosphere at three sites in Australia from 1953 to 1956. In 1957 and 1958 the UK carried out a series of nine atmospheric tests at Christmas Island. The UK conducted its first thermonuclear test in November 1957.

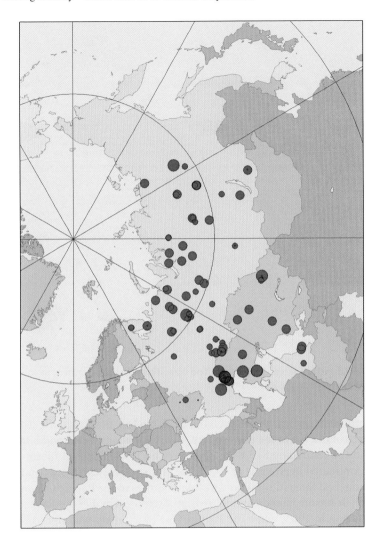

Fig. 1.3 Explosion sites for Peaceful Nuclear Explosions (PNEs) in the former Soviet Union. (Mikhailov 1996, Nordyke 2000). The radius of the symbol is a measure of the size of the explosion

Between 1962 and 1991 the United Kingdom conducted in all 24 test explosions at the Nevada Test Site, in cooperation with the USA.

1.1.8 USA

The USA conducted the first ever nuclear explosion on July 16, 1945, and the USA has also conducted more nuclear tests than any other State. The USA has

Fig. 1.4 Aerial picture from the Nevada Test Site. The subsidence craters in this area are caused by underground nuclear testing

announced 1030 nuclear tests involving 1125 nuclear devices in all. 210 explosions were conducted in the atmosphere. Five explosions were conducted under water and the remaining 815 were underground tests. The first thermonuclear test was conducted in November 1952. The atmospheric tests were conducted at the Nevada Test Site (Fig. 1.4) and at four test sites in the Pacific Ocean: Bikini, Eniwetok, Johnson Islands and Christmas Islands. The USA ended its atmospheric testing in 1963, when the Partial Test Ban Treaty was signed. The first underground test took place in 1957. At times, especially before and after the 1959–1961 moratorium, the USA also had intense test activity with 77 tests in all in 1958 and 96 tests in 1962. During the rest of the 1960s the USA had a test activity of close to 50 tests a year. The last US explosion was carried out in September 1992.

The USA, within its Plowshare program (DOE 2007), also explored the possible benefits of PNEs. 27 explosions, three of which involved multiple nuclear devices, were conducted from 1961 until 1973. The total number of nuclear devices used was 35, some with yields below 100 tons. Many of the tests were related to the development of nuclear explosives for excavation applications. Only three of the PNE experiments were conducted outside the Nevada Test Site. No large-scale American PNE program similar to that of the Soviet Union was ever created.

1.2 Nuclear Weapons – With and Without Testing

1.2.1 Why Testing?

The extensive nuclear testing over half a century is obviously intimately linked to development of nuclear weapons. The first test demonstrated that a nuclear chain reaction in the form of an explosion could be created. Since then, testing has provided detailed insight into the physics of nuclear explosions and how to control their various basic parameters, such as yield, radiation etc. Testing was also used to develop specific nuclear warhead designs to fit operational requirements and different nuclear weapon systems. Series of tests were conducted to develop and validate such new weapons. The development of any complex system, including military and nuclear ones, takes a long time with many years between a first test and the fielding of an operational system. As computer models became available to simulate the explosion phenomena from a given warhead design, tests were used to provide data to calibrate and validate those models. A few tests have also been undertaken to ascertain that warheads in storage function as planned. The earlier testing also studied the effects of nuclear explosions on large numbers of target objects, mainly military. Nuclear tests, especially the very first a State conducts or a particularly powerful one, have also over the years carried great political significance. Tests have been deliberately used by States as an element in their security policy. The large number of tests conducted by the USA and the USSR in the early 1960s and the huge USSR test in 1961 are early examples during the Cold War period. The tests conducted almost simultaneously by India and Pakistan in 1998 and the North Korean test in 2006 are more recent ones.

Even if nuclear testing has proved important for the development of nuclear weapons, there are a few examples where weapons have or are assumed to have been developed without full-scale testing. The nuclear bomb used by the Americans against Hiroshima was an untested gun-type design using uranium. This bomb had a different design and a different nuclear material than the first device tested and the bomb used against Nagasaki, which were both implosion-type plutonium bombs. South Africa has officially acknowledged that it developed and manufactured six gun-type nuclear devices using highly enriched uranium. This development involved a lot of non-nuclear testing, but no nuclear test explosion was carried out. South Africa has also denied any involvement in the mysterious "double flash" observation over the South Atlantic ocean on September 22, 1979 (ISIS 2001). It is generally assumed that Israel has developed and manufactured nuclear weapons. Israeli authorities have neither confirmed nor denied this assumption. No nuclear test explosion that can be associated with a possible nuclear weapons program in Israel has ever been confirmed. There have been speculations that Israel might have been involved, if the September 22 observation came from a nuclear explosion (Wisconsin Project 1996). Unless testing has been conducted in cooperation with any other nuclear weapon State, it has to be assumed that if Israel has developed nuclear weapons it has most likely done so without nuclear testing.

1.2.2 Global Nuclear Capabilities

The discussion of nuclear testing and a test ban is closely linked to issues related to the maintenance and further development of nuclear weapon arsenals. We will therefore briefly summarize the global nuclear capabilities and the plans for the coming years, as publicly known. We base this summary on the UK White Paper on its Nuclear Deterrent (MOD 2006) and on a summary given by the US Arms Control Association (ACA 2007).

In its White Paper the UK Government outlines that the future of the UK nuclear deterrent will be a scaled-down force of three to four new submarines equipped with in all some 50 Trident D5 strategic ballistic missiles. On March 14, 2007 the UK Parliament voted in favor of this renewal of Britain's nuclear weapon system. The Trident missile was deployed in 1992 and is now expected to have its lifetime prolonged until the early 2040s. The total number of warheads will be less than 160 and the design of these warheads will last into the 2020s. The Trident missile and several other strategic ballistic missiles of other nuclear weapon States are capable of carrying multiple warheads.

The UK White Paper notes that China is modernizing its nuclear forces. Its strategic capability currently comprises a silo-based ICBM force of around 20 missiles. China also has a large number of intermediate and medium range ballistic missiles, all believed to carry single nuclear warheads. New projects include mobile ICBMs, ICBMs with multiple warheads and submarine-launched strategic ballistic missiles and potentially nuclear-capable cruise missiles. The US Arms Control Association estimates that China has more than 100 nuclear warheads.

The French nuclear deterrent is now based on two systems; submarine-launched ballistic missiles and air-launched cruise missiles. A new ballistic missile, M51, is under development and will be carried aboard a new class of four submarines to come into service in 2010. France is also developing a new air-launched cruise missile (ASMPA) for deployment on the Rafael aircraft around 2009. Total warhead numbers are around 350.

The UK White Paper also notes that Russia has deployed a triad of land, sea and air based strategic nuclear weapons comprising some 520 intercontinental ballistic missiles, more than 250 submarine-launched ballistic missiles and about 700 air-launched cruise missiles. Under the terms of the Strategic Offensive Reduction Treaty with the USA, Russia, and also the USA, will reduce the number of operationally deployed strategic nuclear warheads to a maximum of 2200 each by 2012. Russia also retains a very large stockpile of non-strategic nuclear weapons. It continues to modernize its nuclear weapons and is deploying a new intercontinental ballistic missile (SS-27) and is testing a new submarine launched ballistic missile. The US Arms Control Association reports that Russia has 4978 strategic warheads, approximately 3500 operational tactical warheads, while in addition more than 11 000 strategic and tactical warheads are stockpiled.

The USA has also a triad of strategic nuclear weapons. The naval component contains 14 Ohio-class submarines each carrying up to 24 Trident D5 missiles, the same as the UK. The silo-based Minuteman intercontinental ballistic missiles are being reduced to 450 and will be sustained until the 2020s. The USA also has air-delivered cruise missiles and free-fall bombs. Similar to Russia, by 2012 the USA will reduce the total number of operationally deployed strategic nuclear warheads to a maximum of 2200 under the Strategic Offensive Reduction Treaty. The USA also has a non-strategic force of nuclear weapons. According to the US Arms Control Association, the USA has 5968 strategic warheads, more than 1000 operational tactical weapons and approximately 3000 reserve strategic and tactical warheads.

Both India and Pakistan have conducted nuclear tests. They are now capable of delivering nuclear weapons by fixed-wing aircraft and land-based ballistic missiles. Development work on warheads and delivery systems continues in both countries. On February 21, 2007 the two countries concluded an agreement aimed at *"reducing the risk from accidents relating to nuclear explosions"* (Wall Street Journal 2007). The US Arms Control Association reports that India has 45–95 nuclear warheads and that the size of the Pakistani stockpile is between 30 and 50 nuclear warheads.

North Korea carried out a nuclear test explosion in October 2006. It has short and medium range ballistic missiles with conventional warheads in service and is developing a long range missile. International efforts have been under way for many years to get North Korea to stop its nuclear weapons ambitions. Under the 12 February, 2007 agreement (BBC 2007) North Korea agreed with the five other nations involved in the negotiations – China, Japan, Russia, South Korea and the USA – to shut down and seal its main reactor and to declare and disable all its nuclear facilities in return for fuel and eventually diplomatic recognition.

Israel is believed to have nuclear weapons capability. It possesses short and intermediate range missiles believed to be capable of delivering nuclear warheads. The US Arms Control Association estimates that Israel possesses between 75 and 200 nuclear warheads.

1.2.3 Nuclear Weapons Without Testing

For many years, both before and during the negotiations and after the signing of the CTBT, there have been many discussions about the consequences of a comprehensive test ban. Will it be possible to maintain existing nuclear stockpiles and ascertain whether they are safe and operational? Will it be possible to replace aging and outdated components with new ones without jeopardizing the proper functioning of the devices? Or do the main nuclear weapon powers have sufficient knowledge to progress with significant weapons development, despite the test ban?

Three of the recognized nuclear weapon States in the NPT – France, Russia and the United Kingdom – have ratified the CTBT and thus decided that they do not need any further testing to support their future nuclear weapons. As for the remaining two – China and the USA – the discussion of these issues has been most public in the USA. Further, as the CTBT was put on the back burner by Washington it is also Washington that has to put it back on the agenda again. Let us therefore look at the American discussion.

In addition to the existing nuclear weapons maintenance program, the USA has created a Stockpile Stewardship Program as a scientific and engineering effort to maintain the US nuclear deterrent in an era of no testing (Jeanloz 2000, DOS 1999). It is to be assumed that the other main nuclear weapon States have similar programs. The Stewardship Program is concerned with methods and procedures of what essentially is non-destructive testing of the "nuclear package" within the device, comprising the nuclear material, the chemical explosives and the ignition elements. The many non-nuclear parts of a nuclear device can be maintained and tested as before. Plutonium is the key component of the primary stage of most nuclear weapons and is a material with complex and variable properties. A lot of effort has gone into developing methods and procedures to diagnose the in-situ status of plutonium inside a "nuclear package". It was long assumed that plutonium would deteriorate as it aged. A study referred to in (Medalia 2007) found that "*there is no degradation in performance of primaries of stockpile systems (i.e. warheads) due to plutonium aging that would be cause for near-term concern regarding their safety and reliability. Most primary types have credible minimum lifetimes in excess of 100 years as regards aging of plutonium.*" An important component of the Stewardship program is the Life Extension Program where aging components are replaced with new ones that replicate the original design and use original materials as far as possible.

Extensive numerical simulation models and corresponding computer facilities have been developed to provide a detailed understanding of the behavior of material during the high temperatures and pressures in a nuclear explosion (Schwitters et al. 2003). A detailed understanding of these conditions is critical if some, even small, changes have to be made to a particular design. The numerical simulations build on experience from previous testing and are calibrated by experimental tests of the actual materials at high pressures and temperatures. Such tests could either be subcritical, involving fissile material in quantities small enough not to sustain a nuclear chain reaction, or hydrodynamic tests, where the fissile material has been replaced with inert material with similar properties. The Stewardship Program also contains an element of basic research intended to maintain the viability of the weapons laboratories. The reliability, safety and security of the US nuclear stockpile has regularly been surveyed, and the assessment reported to the Congress (Foster 2003).

In his assessment in a congressional testimony in 2005, Ambassador Linton Brooks, Administrator of the National Nuclear Security Administration, stated (Brooks 2005) *"today stockpile stewardship is working, we are confident that the stockpile is safe and reliable, and there is no requirement at this time for nuclear tests. Indeed, just last month, the Secretary of Energy and Secretary of Defense reaffirmed this judgement in reporting to the President their ninth annual assessment of the safety and reliability of the U.S. nuclear weapons stockpile... Our assessment derives from ten years of experience with science-based stockpile stewardship, from extensive surveillance, from the use of both experiments and computation, and professional judgment"*.

Despite Ambassador Brooks' assessment, the US administration has decided to reduce the lead-time for resuming tests from 36 to 18 months to be able to respond to unexpected events. This has been criticized, even in the USA. Rep. David Hobson stated in an address at a National Academy of Sciences symposium in 2004 (Hobson 2004) *"I view the efforts to reduce the nuclear test readiness posture to 18 months as very provocative and overly aggressive policies that undermine our moral authority to argue that other nations should forego nuclear weapons. We cannot advocate for non-proliferation around the globe and pursue more useable nuclear weapons options at home. That inconsistency is not lost on anyone in the international community."*

The discussion in the USA on the renewal of nuclear warheads is focused on the concept of the Reliable Replacement Warhead (RRW) Program. The RRW concept is to design a new warhead based on the experience accumulated at the two nuclear weapons laboratories at Los Alamos and Livermore, which are engaged in preparing competing designs, both supported by the Sandia National Laboratories. The design goals for the RRW would include relaxing requirements on efficiency criteria, such as yield and yield-to-weight ratio, increasing warhead security, reliability and longevity, and easy manufacture using more readily available material. The new warhead should be developed and certified without testing and a Nuclear Weapons Complex Infrastructure Task Force reported in July 2005 (Global Security 2005) that *"The Task Force is confident that the Complex* (the nuclear weapons facilities) *can now design a nuclear weapon that is certifiable without the need for underground testing"*. In a statement on April 5, 2006, before the subcommittee on Strategic Forces of the House Armed Service Committee, Mr Thomas D'Agostino noted (D'Agostino 2006) *"Stockpile stewardship is working; the stockpile remains safe and reliable. This assessment is based not on nuclear tests, but on cutting-edge scientific and engineering experiments and analysis, including extensive laboratory and flight tests of warhead components and subsystems."*

1.3 Why a Comprehensive Nuclear-Test-Ban Treaty?

The Preamble to the CTBT (CTBT 1996) gives a summary of the answers, which are both political and technical. The political reason for CTBT is that it *"constitutes an effective measure of nuclear disarmament and non-proliferation in all its aspects"*. Ending nuclear testing is also *"a meaningful step in the realization of a systematic process to achieve nuclear disarmament"*. Nuclear non-proliferation and nuclear disarmament are also contributing *"to the enhancement of international peace and security"*.

The Preamble also identifies some technical reasons for a treaty. A CTBT will be *"constraining the development and qualitative improvements of nuclear weapons and ending the development of advanced new types of nuclear weapons"*. The Preamble also notes *"the views expressed that this Treaty could contribute to the protection of the environment"*.

The Comprehensive Nuclear-Test-Ban Treaty has been on the political agenda for a very long time and it was no doubt a significant political achievement to agree the Treaty. The signing of the Treaty also took place in a hopeful atmosphere. There were expectations that this would mark a new start of international efforts to promote nuclear disarmament and non-proliferation. Many hoped that the CTBT would be followed by a rapidly concluded ban on the production of weapons grade nuclear material, a so-called "Fissile Material Cut-Off Treaty (FMCT)". Hopes faded rapidly when the Conference on Disarmament (CD) proved unable to agree even to start negotiations. The US failure to ratify the CTBT put its entry into force off into an uncertain future. The nuclear tests by India, Pakistan and North Korea and the nuclear dispute with Iran all served to reduce optimism.

So what is left of the political ambitions and expectations expressed in the Preamble? The CTBT has, before entry into force, established a norm of no testing: a norm that it will be politically very costly for any State to break. It has thus in practice put an end to nuclear testing without formally being in force. The cooperation within the Preparatory Commission of the Comprehensive Test-Ban-Treaty Organization, responsible for the implementation of the treaty and for the establishment of the most extensive verification regime ever created, has in and by itself been a significant confidence building measure. Signatory States, with few exceptions, have demonstrated good political will to engage in the build-up of this complex, intrusive verification regime. This is certainly a political achievement.

On the technical side, would the CTBT have any effect on constraining the development of new or improved weapons and preventing new nuclear weapon States? The established nuclear weapon States have large programs in place to maintain and to keep their nuclear weapons up-to date. It is to be expected that those programs, discussed above, will provide tools to also upgrade and modernize the weapons including the "nuclear package". Nuclear weapon States, including those with experience from extensive past test programs, may or may not feel confident in developing and fielding completely new nuclear weapons designs without testing. It is reasonable to assume that a test ban has a restraining effect on ambitions to design new "nuclear package" concepts.

South Africa and most likely Israel have demonstrated that it is possible to develop nuclear weapons without testing. The South African weapons were, as discussed above, based on a well known, proven design. The North Korean explosion showed that even a country with most limited resources can develop a nuclear device, imperfect as it might have been. It is thus reasonable to assume that any country with access to the necessary amount of fissile material and some technical and scientific expertise will be able to develop a primitive nuclear device. To develop a "nuclear package" to a qualified weapons system such as long range missiles is quite another challenge. Here it is most likely that testing would be needed or at least highly desirable to provide confidence in the system. A CTBT thus imposes, on most non-nuclear weapon States, a strong limitation on the development of an advanced operational nuclear weapon system.

1.4 Related Nuclear Arms Control and Disarmament Treaties

The nuclear Non-Proliferation Treaty (NPT) (IAEA 1970) is often referred to as the key guardian against nuclear proliferation. The Treaty was signed in 1968 and entered into force in 1970. The NPT aims at stopping the spread of nuclear weapons beyond the five States that possessed nuclear weapons on 1 January 1967: China, France, UK, USA and USSR. The nuclear weapon States undertake "*not to transfer to any recipient whatsoever nuclear weapons or other nuclear explosive devices or control over such weapons*". The nuclear weapon States further undertake to "*not in any way assist, encourage or induce any non-nuclear-weapon State to manufacture or otherwise acquire nuclear weapons*". The non-nuclear-weapon States undertake not to receive any nuclear weapons or manufacture or in any way acquire such weapons. Each non-nuclear weapon State undertakes to accept safeguards established with the IAEA to verify its obligation under the NPT. The safeguards have been extended by the introduction in 1997 of the so-called "Additional protocols" (IAEA 2007a). The NPT should not "*affect the inalienable right of all the Parties to develop research, production and use of nuclear energy for peaceful purpose without discrimination*". In the preamble to the Treaty the Parties declare "*their intention to achieve at the earliest possible date the cessation of the nuclear arms race and to undertake effective measures in the direction of nuclear disarmament*". NPT article VI also includes a commitment by all Parties, and this is directed primarily to the nuclear weapon States, "*to pursue negotiations in good faith on effective measures relating to cessation of the nuclear arms race at an early date and to nuclear disarmament.*"

NPT is a most important treaty but also a treaty in trouble. It is a treaty designed to maintain the status quo when it comes to nuclear weapon States, yet three States – India and Pakistan, which have demonstrated that they have nuclear devices, and Israel, which is generally believed to have nuclear weapons – are not parties to the Treaty. This creates the serious question of how to handle these States in relation to the Treaty, as there is no way of fitting them into the existing framework. They were not nuclear weapon States in 1967, which is the NPT definition of a nuclear weapon State, but they possess nuclear devices today. North Korea, originally a party to the NPT, is in a similar situation. This is only one of the problems. The Treaty contains two kinds of commitments: for the non-nuclear weapon States not to acquire nuclear weapons and for the nuclear weapon States to work actively towards nuclear disarmament. Many States are not satisfied by the nuclear weapons States' action so far in this regard and they hold the view that the nuclear weapon States are not living up to their NPT commitments. These serious concerns have turned the NPT Review Conferences into most difficult meetings. These conferences, intended to boost the Treaty, have instead placed it in jeopardy. There is a third problem related to the right of all NPT Parties to benefit fully from the use of nuclear energy for peaceful purposes as long as they fully comply with the verification provisions established with the IAEA. The Iranian ambition to create all the elements of the fuel cycle nationally has caused international concern. Even if all activities are to be placed under IAEA control there are still concerns that such a program provides a basis that might be used for a nuclear weapons program. This is an example that treaties operate in a changing political environment and that mutual trust and confidence are important prerequisites for a functioning treaty regime.

There are three treaties that ban nuclear weapons in specific environments that are our common heritage. The first treaty to be signed was the *Antarctic Treaty* (British Antarctic Survey 2007) which entered into force in 1961. This treaty stipulates that Antarctica shall be used for peaceful purposes only. All military activities, including the presence of any kind of weapons, are prohibited. The *Outer Space Treaty* (US Department of State 2007) of 1967 prohibits the placing of all kinds of weapons of mass destruction in orbit around the earth. The *Sea-Bed Treaty* (Atomic Archive 2007b), which entered into force in 1972, similarly prohibits the emplacement of any nuclear weapon or any other weapon of mass destruction on the sea bed, the ocean floor and the subsoil thereof. The Treaty covers areas beyond the outer limit of a twelve-mile zone defined as Territorial Sea (Convention on the Territorial Sea and Contiguous Zone 1958).

A number of regional treaties have been established that ban nuclear weapons, including their testing, in several parts of the world. The first such treaty was the *Latin America Nuclear Free Zone Treaty*, better known as the *Treaty of Tlatelolco* (IAEA 2007b). Tlatelolco is a section of Mexico City where the Treaty was signed in 1967. It entered into force one year later and has been adhered to by 33 Latin American States making the whole of Latin America a nuclear-weapon-free zone. The *South Pacific Nuclear-Free Zone Treaty,* also referred to as the *Treaty of Rarotonga* (CNS 2007a), was signed in 1985 and entered into force a year later. The treaty zone covers an extensive part of the South Pacific. The Treaty has 13 permanent members from South Pacific and the protocols are signed by the five nuclear weapon States. Southeast Asia is also covered by a similar treaty, the *Southeast Asia Nuclear-Weapon-Free Zone Treaty (SEANWFZ)* or the *Treaty of Bangkok* (CNS 2007b). The Treaty has 10 members and entered into force in 1997. As early as 1964, the Organization of African Unity (OAU) adopted a declaration on the denuclearization of Africa. In 1996 this declared intention was turned into the *African Nuclear-Weapon-Free Zone Treaty* or the *Pelindaba Treaty* (CNS 2007c). The Treaty covers all of the African continent, island members of OAU and islands considered by the OAU to be part of Africa. The Treaty prohibits any nuclear testing and any other activity related to nuclear weapons. It also covers some aspects of civilian nuclear activities and prohibits the dumping of radioactive wastes.

Box 1
What Is a Nuclear Device?

The first nuclear weapon was developed during the Manhattan Project (Fig. B1.1) (Condon 1943, Serber 1992). The fundamental difference between a nuclear and a chemical reaction is that a nuclear reaction involves the nuclei but a chemical reaction only the electrons of an atom. In a chemical process the elements are the same after a reaction as before, although they may appear in new combinations. A nuclear reaction creates new elements. There are two principally different ways of releasing nuclear energy: *fission,* where heavy elements are split into lighter ones, and *fusion,* where light elements are combined into heavier ones. In both cases the masses of the created elements are less than those of the original ones. This mass difference (m) is released as energy (E) following Einstein's famous formula $E = mc^2$, where c is the velocity of light. Both fission and fusion processes are used in nuclear weapons.

Uranium (U) and plutonium (Pu) are the two materials used in fission weapons (FAS 2007). Only specific isotopes of these materials can be used: U-235, U-233 and Pu-239. The uranium isotopes exist in nature in concentrations below 1% and are separated from the main isotope U-238 by either filtering in large-scale gas-diffusion plants or in gas centrifuges where the lighter and heavier elements are separated by cascades of thousands of high speed centrifuges. Pu-239 is produced from U-238 in a nuclear reactor. At the same time, another plutonium isotope, Pu-240, which is unsuitable for weapon application, is produced in concentrations that increase over time. To obtain Pu-239 with the necessary purity of about 93% it is necessary to irradiate the uranium only for a short time, much shorter than in normal power reactors. Weapons-grade plutonium is thus normally produced in specialized reactors. The plutonium is chemically separated from the uranium fuel in nuclear reprocessing plants.

A nuclear explosion is a rapid chain reaction in which a uranium or plutonium nucleus captures a neutron and then splits, generating energy and new neutrons (Fig. B1.2). A minimum amount of fissile material is needed to sustain the reaction by providing the necessary neutron flux. This *critical mass* depends on the material used and its purity, the geometry of the nuclear device, and the use of reflectors to mirror neutrons back into the process. Two principal designs for creating a critical mass are the gun-type, where two pieces of fissile material are accelerated – shot – into each other using conventional explosives

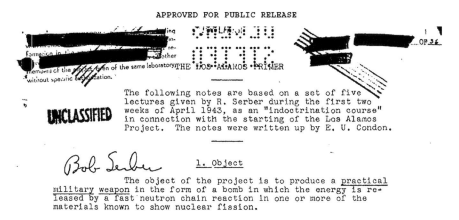

THE LOS ALAMOS PRIMER

The following notes are based on a set of five lectures given by R. Serber during the first two weeks of April 1943, as an "indoctrination course" in connection with the starting of the Los Alamos Project. The notes were written up by E. U. Condon.

UNCLASSIFIED

1. Object

The object of the project is to produce a <u>practical</u> <u>military weapon</u> in the form of a bomb in which the energy is re-leased by a fast neutron chain reaction in one or more of the materials known to show nuclear fission.

Fig. B1.1 Notes in "The Los Alamos Primer" stating the objective of the Manhattan Project

(Fig. B1.3), and the implosion-type, where a sphere of fissile material is compressed, also by conventional explosives. The gun-type with uranium as fissile material was used for the bomb dropped on Hiroshima (Fig. B1.4b), while the implosion technique and plutonium were used in "Trinity" (Fig. B1.4a), the first nuclear test, and in the Nagasaki bomb (Fig. B1.4c). A gun-type design is shown in Fig. B1.5a. Critical masses for bare spherical assemblies are reported to be some 10 kg for Pu-239, 15 kg for U-233 and about 50 kg for U-235. In real charges reflectors and tampers are used which reduce these numbers by a factor of three (Sublette 2007a). For safeguard purposes, the IAEA defines one significant quantity of fissile material as 8 kg of Pu-239 or U-233 or 25 kg of U-235 (Zuccaro-Labellarte and Fagerholm 1996).

The explosive yield of a nuclear fission device depends on the amount of fissile material and on the actual design. The yield of a quite simple fission device with a critical mass of fissile material will be in the range 10 000–20 000 tons (10–20 kt). The first device exploded by the nuclear weapons States all had yields in that range. The yield of the explosion conducted by North Korea on

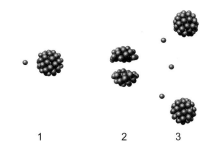

1   2   3

Fig. B1.2 Nuclear chain reaction. 1: A neutron is about to hit the uranium nucleus. 2: The nucleus splits (fission) into several smaller atoms, releasing radiation and several more neutrons. 3: The chain reaction begins; released neutrons hit other nuclei causing them to split and so on

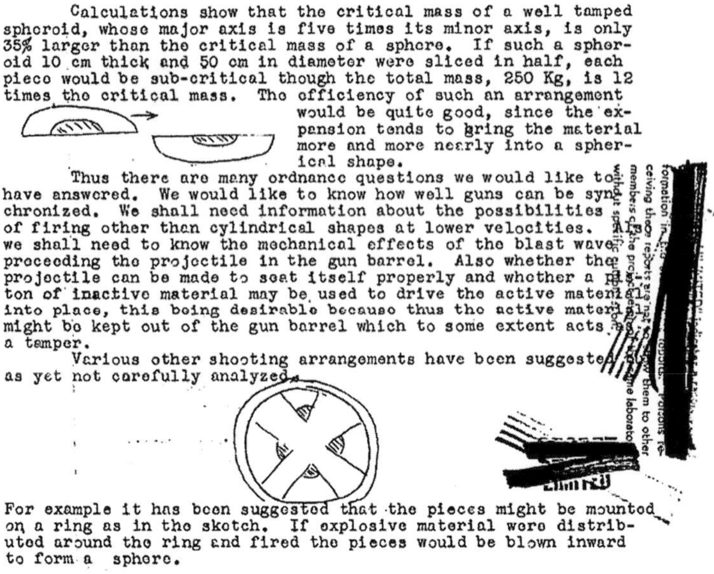

Calculations show that the critical mass of a well tamped
spheroid, whose major axis is five times its minor axis, is only
35% larger than the critical mass of a sphere. If such a spheroid 10 cm thick and 50 cm in diameter were sliced in half, each
piece would be sub-critical though the total mass, 250 Kg, is 12
times the critical mass. The efficiency of such an arrangement
would be quite good, since the ex-
pansion tends to bring the material
more and more nearly into a spherical shape.

Thus there are many ordnance questions we would like to
have answered. We would like to know how well guns can be syn-
chronized. We shall need information about the possibilities
of firing other than cylindrical shapes at lower velocities. Also
we shall need to know the mechanical effects of the blast wave
proceeding the projectile in the gun barrel. Also whether the
projectile can be made to seat itself properly and whether a pis-
ton of inactive material may be used to drive the active material
into place, this being desirable because thus the active material
might be kept out of the gun barrel which to some extent acts as
a tamper.

Various other shooting arrangements have been suggested
as yet not carefully analyzed.

For example it has been suggested that the pieces might be mounted
on a ring as in the sketch. If explosive material were distributed around the ring and fired the pieces would be blown inward
to form a sphere.

Fig. B1.3 Early designs of a fission bomb from "The Los Alamos Primer"

Fig. B1.4a,b,c (continued)

Fig. B1.4a,b,c The first three nuclear explosions: (**a**) "Trinity" on July 16, 1945; (**b**) Hiroshima August 6, 1945; and (**c**) Nagasaki August 9, 1945

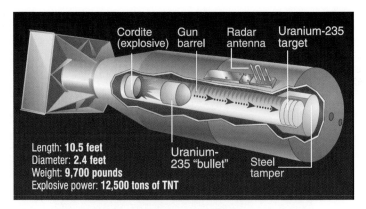

Fig. B1.5a Design of a "gun type" nuclear weapon. This design is called "Little Boy". The uranium-235 bomb that destroyed Hiroshima was flown there in a B29 bomber and aimed with a bombsight. When the bomb fell to 1900 feet, a radar antenna set off a conventional explosive in the bomb chamber. This catapulted an uranium-235 wedge through the gun barrel into the uranium-235 target rings, producing a self-sustaining nuclear chain reaction. From Cox (1999)

October 9, 2006 was a factor of ten smaller. Devices can be made to produce lower yields by a more sophisticated design and weapons with yields below 1 kt have been designed. Yields can also be reduced if there are some problems with the material or with the functioning of the device. More fissile material can be added to increase the yield.

Another, more efficient method is to add fusion material and create a fusion or thermonuclear reaction. The fusion materials used are two different isotopes

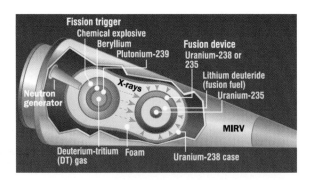

Fig. B1.5b Design of a modern thermonuclear weapon. This W87 thermonuclear warhead is launched on an MX intercontinental missile. Packed into a multiple independently targeted re-entry vehicle (MIRV), it splits off from the missile to strike its target. The MIRV length is 5.7 feet and base diameter is 1.8 feet. The explosive power is 300 kt. The compression of plutonium with a chemical explosive starts a fission explosion that in turn is boosted by the fusion of the deuterium-tritium gas. X-rays then compress the second component, causing a larger fission/fusion. From US News and World Report 31 July 1995 and Cox (1999)

Fig. B1.6 Atmospheric thermonuclear test explosion "Truckee" with estimated yield of 210 kt, conducted by the USA on 9 June 1962 south of Christmas Island

of hydrogen; deuterium and tritium, having one proton and one and two nucleons, respectively. Deuterium exists in low concentrations in natural hydrogen, from which it can be enriched. Most of the tritium is produced by lithium in thermonuclear fuel, when it is exposed to high fluxes of neutrons. To initiate a fusion reaction an initial fission reaction must be used to create the temperatures and pressures needed to initiate a thermonuclear explosion. In the so-called Teller-Ulam configuration (Sublette 2007b) the X-ray flux from the primary fission charge is directed onto the surface of a separate secondary package where it creates an enormous pressure in the thermonuclear fuel, generating conditions for an effective thermonuclear chain reaction. Such devices, usually referred to as hydrogen bombs, can have yields corresponding to millions of tons (Mt) of conventional explosives (Figs. B1.5b and B1.6). The largest explosion ever carried out was a Soviet hydrogen bomb with an estimated yield close to 60 Mt.

Chapter 2
Monitoring Technologies

The development and testing of nuclear weapons was paralleled by the development and implementation of monitoring techniques and technologies. The purpose was initially intelligence gathering, and during the period of atmospheric testing a lot of information could be obtained from the analysis of radioactive fallout. Systems were also established in some countries to detect a nuclear weapons attack. In the first chapter we have given the background to nuclear testing, nuclear weapons and international treaties related to such weapons. The purpose of this chapter is to give a corresponding background to the methods and techniques that have been developed over the years to monitor nuclear testing and that eventually became part of the verification regime of the CTBT. In giving this background we will make some references to the International Monitoring System (IMS), which is a key part of the CTBT verification regime and which is discussed from different perspectives in Chapters 4, 6 and 7.

2.1 Nuclear Explosions – Detectable Features

A nuclear explosion generates a number of features, or signals, that may be detected, some at large distances from the source, others at close distances only. The signals differ if an explosion is carried out in the atmosphere, under water or underground. An explosion in the atmosphere immediately generates a strong, characteristic "double flash" (Schmidt 2007) light pulse, an electromagnetic pulse and a burst of ionizing radiation. A nuclear explosion also generates radioactive gases and particles, emanating from the exploding device and from the interaction of the primary radiation with the construction materials and the environment. This radioactive fallout can be carried in the atmosphere to great distances. The explosion shock wave generates a low frequency signal in the atmosphere, a so-called infrasound signal. When the shock wave hits the ground a seismic signal is generated. Explosions in the atmosphere over land also generate a significant, lasting environmental impact, destroying any vegetation or structures within a large area and inducing radioactivity in the ground (Glasstone and Dolan 1983).

O. Dahlman et al., *Nuclear Test Ban*, DOI 10.1007/978-1-4020-6885-0_2,
© Springer Science+Business Media B.V. 2009

Underwater explosions in deep ocean waters generate a very strong hydroacoustic signal. Explosions in water also generate comparatively strong seismic signals due to the good coupling between the shock wave in the water and in the sub-surface bedrock. As an underwater explosion is likely to create a water cascade and break into the air it is also likely to generate an infrasound signal. The ground zero effects of an underwater explosion fade away quite rapidly due to the movement of the water. The radioactive material is likely to be washed out locally by the water cascade, which would reduce fallout at great distances. Carrying out a test in rainy weather would further enhance this effect. Gases like the radioactive noble gases produced in the explosion will, however, remain airborne and can be detected far away downwind.

An underground explosion creates a complex dynamic process with the surrounding bedrock (Sublette 2001). Material is melted, crushed and displaced and a cavity is generated, its size and shape depending on the bedrock, the yield and the depth of the explosion. From a fully contained explosion the only effect that can be observed from a distance is seismic signals. The strength and other characteristics of these signals depend on the yield, the local geological conditions and the transmission properties between the explosion site and observing stations. Close-proximity effects of an underground explosion might include leakage of radioactive noble gases through cracks in the ground. This radioactivity might be observed during an on-site inspection, but it might also be observed at large distances. Minor surface effects, including subsidence craters, might be observed. The underground cavity could be detected through geophysical measurements and the slow deterioration of the cavity may create cracking that later can be observed by close-in seismometers.

Evasion scenarios, for which signals from underground nuclear explosions are reduced or mixed with those of nearby earthquakes, have been discussed earlier, for example in the study by the US National Academy of Sciences (NAS 2002), and a summary is also given by Dahlman and Israelsson (1977). The most frequently discussed scenario is the detonation of a small nuclear explosion in a very large cavity to reduce the strength of the seismic signals. This has been tested using chemical charges and a sub-kiloton nuclear device (Springer et al. 1968). Large cavities are needed to reduce the signals even from very low yield explosions. The increased risk of radioactive leakage from explosions in cavities, the logistical problems of creating the large cavities needed, and the probability that the large-scale construction work would be detected by other means, in particular by today's satellite-based methods, have reduced concern that States might use this method for evasive testing.

2.2 Detection, Location and Identification

The capability of any monitoring system is defined by its ability to detect, locate and identify events of interest. In this section we describe how signals from seismic, hydroacoustic and infrasound events are detected and how such events are defined, located and identified. The intent is to also give a background to detection and identification probabilities and event location accuracies. The principal issues apply also to radionuclide detection, but the methods and procedures are quite different and are discussed in the section on radionuclide monitoring later in this chapter.

2.2.1 Detection

Detection is the process of finding a signal or a feature in the ever-present noise. Detection capability depends on the properties of the signals and the noise and on our ability to exploit differences in those properties (Heeger 1997).

2.2.1.1 Signals

We define a signal as a feature generated by an event of interest, be it a nuclear explosion, an earthquake or any other event. A signal has a number of properties. It has a time history and can be short and pulse-like or spread out over a longer time. It has a particular frequency content or a signal spectrum that can also vary substantially. A signal also arrives at a certain time and from a particular direction with a certain apparent velocity. All these characteristics are used when detecting, locating and identifying an event.

In the cases of seismic and hydroacoustic signals the characteristics depend on the source but also to a large extent on the properties of the medium through which the signal is propagating. The earth and, to a lesser extent the oceans, are complex media with great variations in transmission properties both laterally and with depth. Signals from one and the same event observed at different stations can be quite different, even if the stations are at similar distances from the source. On the other hand, conditions are stable over time and signals at one and the same station from events of similar nature, originating close to each other, normally exhibit a similar appearance and properties. For a given station it is thus possible to optimize the detection process for events in a particular region using this past experience.

For infrasound signals the constantly changing atmospheric conditions have a dominant influence on the signals. The influence of the source is hardly discernible. As the atmospheric conditions change over time, so do the signal transmission properties. This means that signals observed at a given station from events in the same region are likely to be very different when the events occur at different times.

2.2.1.2 Noise

Signals are always detected against the background noise, which places an ultimate limit on our observations. Noise can have different origins. For some observations the noise generated by the instrument used to detect the signal might be significant, and become a limiting factor. This is no longer the case for the observations we describe here, where modern instruments have internal noise well below the background noise in the earth, the oceans or the atmosphere. It is actually variations in the atmosphere, such as wind, turbulence and pressure changes, that cause most of the noise not only in the infrasound but also in the seismic and hydroacoustic observations. Other sources are different human activities, like traffic and heavy machinery operating in the neighbourhood of the instrument. Noise is also generated by events too weak to be defined as specific events.

Noise, like signals, has a number of properties. It varies from site to site. Seismic stations inside large continents and away from human activities generally have significantly lower levels of noise than sites close to oceans. For infrasound, in particular, the noise variations over time are most significant, but hydroacoustic and seismic noise also show considerable variation over time and with changing seasons and weather.

Noise has certain spectral characteristics. Seismic noise has been studied in great detail all over the globe and there is a general trend, observed globally, that noise levels decrease with increasing frequency. The spectral composition is site-dependent and could vary significantly from one part of the world to another. Part of the noise can be correlated over short distances of one or a few kilometers. At greater distances, such as between stations in the monitoring network of the CTBT, the noise is uncorrelated.

2.2.1.3 Detection Processes

The traditional, most obvious way of detecting a signal is to look at a recording (for example a seismogram from a seismic station) to recognize a change in the waveform. The human brain is a very good detector. The possibility of detecting a signal depends on the signal-to-noise ratio (SNR). This ratio can be enhanced by filtering the recorded data into a frequency band with a high SNR for the signal of particular interest (Fig. 2.1). Frequency filtering is thus one important tool used in the detection process.

Automatic detection procedures are being used to cope with the large amount of data in regional and global networks. For stations with only one sensor the detection process normally involves filtering the incoming data in a frequency band that gives the highest SNR for the signals of interest. Optimizing this process requires different filtering for signals from events occurring in different regions. An automatic detection algorithm then usually compares the short-term variation of the data to the long-term variation, in the same way as is described for an array station (see Box 2). In this way the detector "floats" on the noise and adjusts the detection capability to the actual noise, giving higher capability when noise is low and vice versa.

Modern monitoring stations, including key seismic and all infrasound stations in the monitoring network of the CTBT, are array stations having a

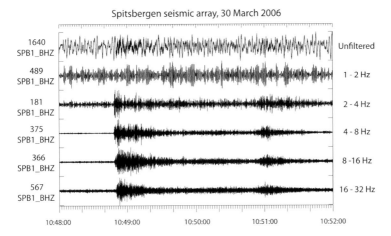

Fig. 2.1 The effect of frequency filtering seismic data. The *upper trace* shows the recording at one element (SPB1_BHZ) of the IMS auxiliary seismic array station located in Spitsbergen, Norway from a small seismic event (magnitude 2.3) near Novaya Zemlya, at a distance of 1300 km. The other traces are the results of filtering this trace in various frequency bands. No signals are visible in the 1–2 Hz band. The signal-to-noise ratio is seen to be highest in the 4–8 Hz band. Numbers to the left represent maximum amplitudes for each trace

number of elements, usually deployed in a symmetric pattern to form an antenna. A modern seismic array may have as many as 25 seismometers placed in a circular pattern with a diameter of a couple of kilometers or more. As is discussed in the Box on array stations, such an array station will improve the SNR by almost a factor of five and also provide preliminary locations of detected events. The corresponding dimension of an infrasound array is of the order of a few hundred meters up to about 2–3 km.

In the detection process there is always a risk of mistaking noise for a signal. The probability of such false detections increases when the detection threshold is decreased to detect weaker signals. This trade-off between the probability of detecting a signal of interest and the probability of a false alarm is usually referred to as a Receiver Operating Curve (Heeger 1997). In any system a judgment must be made on the number of false alarms that are acceptable. For a given situation this also defines the detection threshold. If you are prepared to accept more false alarms, you increase your ability to detect weak signals and vice versa.

The detection capability of a particular network, such as the IMS, depends on a combination of the capabilities at individual stations, and network detection criteria have to be decided. One such criterion for the network to define an event is the minimum number of stations that need to detect a signal. In the same way as for a single station, the capability of a network is a trade-off between detection capability and false alarms. Computer programs have been

Fig. 2.2 Detection probability as a function of the body wave magnitude m_b, for a single short period seismic sensor of the Hagfors array in Sweden. After Dahlman and Israelsson (1977)

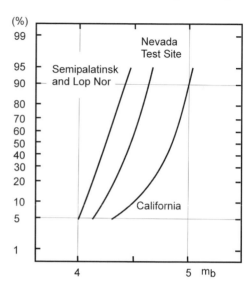

developed to show on-line the current detection capability of a network based on the actual station data. This is discussed further in Chapter 7. As the detections from many stations are combined, each station can operate at a fairly high false alarm rate without creating a large amount of false alarms for the network as a whole. Optimizing the capability of an integrated network is a delicate task that requires considerable testing to fine-tune and balance the detection parameters at individual stations and in the network processing.

Detection probabilities may be looked at from different perspectives. If you are on the monitoring side you may want a high detection probability of say 90% to be quite sure that you are not missing important events. If, on the other hand, you are considering conducting a clandestine nuclear test, you are most likely prepared to take only a small risk of a few percent of being detected. As Fig. 2.2 shows, the difference between 90 and 5% detection probability for a typical seismic station is almost a factor of 3 in signal strength. These two perspectives are essential when we later discuss the capabilities of the CTBT verification system. The deterrence levels can thus be considerably below the levels considered to provide confident detection.

2.2.2 Event Location

In addition to signal detection, an array station also gives a rough location of the event that generated the signal. This is valuable information when facing the next step: to determine which detections at various stations of the network can be associated with one and the same event. For large seismic events observed at many stations this is a fairly straight forward process. For small events, observed at a few stations only, a complex computer trial-and-error process is needed to identify which observations are likely to originate from the same event. This process uses certain criteria, such as the number of detecting stations required to define an event. Once an initial group of stations is established that is assumed to have observed the same event, a preliminary location is estimated. This is a three-dimensional, iterative computational process, minimizing the differences between the observed signal arrival times at the stations and those estimated from an assumed source location and time of the event's origin. The process moves the trial source location around to find this minimum. Station observations can be deleted and added into the process as it progresses. Another approach to event location, now widely used in the seismological community, assumes that events are occurring regularly in time on a global grid system. For each of these hypothetical events, the expected signal arrival times are calculated for all stations of the network. These estimated arrival times are checked against those of actual signal detections, and when there is a sufficient number of matches, an event is declared. The location of the event and its time of origin are subsequently refined.

Locations of events are estimated in three dimensions: latitude, longitude and depth or altitude. The time of origin of the event is an additional parameter. Due to the geometry of the situation, the uncertainty in depth or altitude estimation is normally much greater than in latitude and longitude. For infrasound events the uncertainty in altitude estimates is of the order of 10 km. As the oceans are shallow in relation to the globe as a whole, the location process will not be able to provide a depth for an event occurring in the ocean. For seismic events, too, the depth estimation is difficult and depends on the station distribution around the event; stations close to the event improve the depth estimation capability. Normally only events that occur at depths of 50–100 km can be confidently located in all three dimensions. If so-called depth phases, which are signals reflected at the earth's surface above the hypocenter of an event, can be observed, the depth estimates can be significantly improved.

In the process of associating observations and locating events a model is needed for the velocity distributions of the media through which the observed signals have propagated. The solid earth has a complex velocity distribution with significant variations on local and regional scales, but the velocities do not change over time. It is thus possible, once and for all, to calibrate the earth, a time-consuming undertaking that to a large extent has been carried out, as we shall discuss in the section on seismological monitoring. The velocity distribution of hydroacoustic waves is fairly homogeneous across the oceans and also quite stable over time, with a small variation with ocean temperature. The velocity distribution of infrasound waves is critically dependent on the actual meteorological situation and the value of calibration will gradually fade as the meteorological situation changes. The uncertainty in the velocity distribution of the infrasound waves is thus much greater than for propagation of seismic and hydroacoustic waves.

Using such arguments, perhaps with different emphasis by different States, only a few places on earth would be of interest for close monitoring. The CTBTO and its international verification system must, however, be even-handed and provide whatever data it may obtain on a global basis. The Treaty protocol states that among the products to be provided should be "*Standard screened event bulletins ... screening out events considered to be consistent with natural phenomena or non-nuclear man-made phenomena*". Annex 2 to the protocol lists a number of tentative parameters to be used in this screening process and it is also recognized that this is still very much work in progress and that these procedures are to be further developed as more experience is gained from the IMS.

The "*non-nuclear man-made events*" referred to are large-scale conventional explosions. As there is no way, without an on-site inspection, to distinguish a fully contained nuclear explosion from a non-nuclear explosion fired as a single charge, the Treaty has provisions for voluntary notification of any non-nuclear explosion of 300 tons of explosives or greater, detonated as a single charge. In addition, the Treaty has provisions for voluntary contribution of information on all other chemical explosions greater than 300 tons.

The problem of natural phenomena is easy to formulate. For seismic events it reads: is the event we just detected and located a typical earthquake from that particular region? It is more difficult to provide an answer as the earthquake sources also vary within a limited earthquake region, in addition to the variations in transmission properties to the monitoring stations, all of which significantly affect the observed signals. The dramatic development in data analysis methods and procedures over recent decades, discussed below, has provided tools that might prove most useful in addressing this issue. This, together with the increasing volume of data now being made available as the CTBT global monitoring system is approaching completion, may soon make it possible to address, and provide an answer to, the question: is this an observation from just another earthquake in the region?

2.2.4 New Developments

Most of today's procedures for analyzing data from seismic networks have in principle remained unchanged for decades. The technology in all parts of the system, on the other hand, has developed dramatically during the last two decades. Station instrumentation has improved, additional array stations with improved detection capabilities and the ability to provide initial event locations have been established, a global communications system makes it possible to transmit all information from all stations to a data center where the storage and computational capacity is available to handle all that data. Also, more generally, methods and procedures to handle and analyze very large volumes of data have developed dramatically over the same period of time, as we discuss towards the end of this chapter. How are we benefiting from this increased capability in our analysis? Have we developed the analysis methods and the computational procedures correspondingly? The answer is no.

In the process of detecting and associating observations we are only marginally utilizing the vast amount of information that already has been accumulated in the system on observations of past events in all parts of the world. We are likely to significantly improve our ability to detect and locate events by utilizing a substantial part of the recorded waveforms, and not only a few extracted parameters, and compare those waveforms with earlier observations. This procedure, referred to as "precision seismology" (Richards 2004), which makes use of so-called cross-correlation methods, has been demonstrated in research projects. The full use of all available observations from a particular event, and of stored information and observations of past events, should make it possible to provide a better understanding of the event. This should result in a bulletin that provides a better basis for States Parties to interpret observed events.

2.3 Seismological Monitoring

Seismology is a mature science, which plays an important role in mitigating earthquake hazards and in understanding the earth's interior. A large number of seismological institutes exist and several thousands of seismological stations are operating around the globe. Viable research work is carried out at many places and seismological expertise is available worldwide. There are many excellent scholarly textbooks on seismology and earthquakes (Richter 1958, Shearer 1999, Bolt 1999, Lay and Wallace 1995, Stein and Wysession 2003, Dahlen and Tromp 1998, Aki and Richards 2002).

2.3.1 Seismology and Earthquakes

For most people, seismology is related to the occurrence of earthquakes and tragic disasters. Indeed, more than 100 000 earthquakes are recorded every year. Earthquakes are generally created by large scale, global dynamic processes where a small number of huge plates move in relation to each other. Using a variety of space-based geodetic technologies (NASA 2007), it is now possible to measure these motions, varying from a few mm/year up to more than 50 mm/year. The boundaries of the plates are marked by the frequent occurrence of earthquakes, forming a most consistent pattern (Fig. 2.5).

An earthquake is a motion along a boundary or a fault, starting when the stress exceeds the strength of the rock material. To simplify, it is rather like bending a stick until it breaks. A huge earthquake under the ocean off the cost of Sumatra on December 26, 2004, had a fault line of 1200 km, along which the

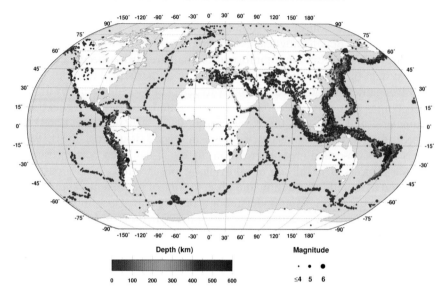

Fig. 2.5 Global distribution of 27 574 seismic events, mainly earthquakes, contained in the Reviewed Event Bulletin of the International Data Center (IDC) for 2006. See Chapter 6 for an account of the IDC and its products

sea bottom was displaced vertically by as much as 15 m. This is one of the largest earthquakes ever observed and measured 9 on the magnitude scale. Magnitude is a logarithmic measure of the amplitude of the recorded signal adjusted for the distance between the station and the event. Events with a magnitude as small as 0, or even negative numbers, can be observed in the vicinity of a station. Earthquakes could thus differ in size by more than 9 orders of magnitude, corresponding to the relation between 1 mm and 1000 km. Table 2.1 provides frequency-magnitude statistics for earthquakes of magnitude 2 and higher. Normally an earthquake needs to have a magnitude of about 2 to be felt by man.

Table 2.1 Frequency-magnitude statistics for earthquakes worldwide

Descriptor	Magnitude	Average annual number
Great	8 and higher	1[1]
Major	7–7.9	17[2]
Strong	6–6.9	134[2]
Moderate	5–5.9	1319[2]
Light	4–4.9	13 000 (estimated)
Minor	3–3.9	130 000 (estimated)
Very Minor	2–2.9	1 300 000 (estimated)

[1] based on observations since 1900, [2] based on observations since 1990
Data from USGS (NEIC/USGS 2007)

A number of earthquake disasters occur every year. The Sumatra earthquake on December 26, 2004 took the lives of a quarter of a million people (Fig. 2.6). Most of them died in the huge tsunami generated by the earthquake. A tsunami is a wave that propagates with the speed of a commercial jetliner and with a fairly low amplitude in the deep ocean, dramatically increasing its amplitude when it approaches shallow coastlines. A Chinese earthquake in 1556 is considered to be the most devastating ever, taking more than 800 000 lives. An earthquake in the Tangshan province of China in 1976 killed some 250 000 people. Another well-known, disastrous earthquake is the one in Lisbon in 1755, with an estimated magnitude in the range 8.6–8.7, also creating a tsunami and killing about 100 000 people. The San Francisco earthquake in 1906 is also famous for the huge fires that were created. Its death toll was about 3000.

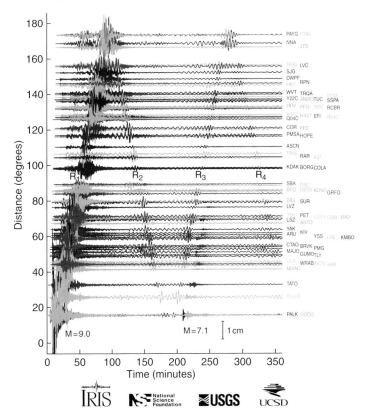

Fig. 2.6 Recordings from stations of the IRIS Global Seismograph Network of the large Sumatra earthquake on 26 December 2004. The surface waves (R1, R2, R3, ...) of this earthquake circle the earth in multiple orbits

2.3.2 Seismic Waves and Wave Propagation

An earthquake or an underground explosion generates several kinds of seismic signals. These signals are normally divided into body waves, propagating through the interior of the earth, and surface waves that are confined to the earth's surface. The body waves are of two different kinds: P-waves or primary waves, which are acoustic waves, and S waves or shear waves for which the oscillations are perpendicular to the propagation direction of the wave. P wave velocities are about 1.7 times the velocities of the S waves. The P waves are thus the first to arrive at a station and they are also of prime importance in seismic monitoring. P waves are the key waves used for the routine detection and location of seismic events, and they are used together with S waves and surface waves for event identification.

There are also two main kinds of surface waves: Rayleigh and Love waves propagating along the earth's surface. Both waves have dominant signal periods in the range 10–100 s and are dispersive, meaning that the velocity is frequency dependent, with the lowest frequencies having the highest velocity. Velocities of surface waves lie in the range 3–4 km/s, depending on the travel path. Explosions generally generate weaker surface waves than earthquakes and a comparison of the amplitudes of P and surface waves has been widely used to discriminate between explosions and earthquakes.

P-waves are thus the waves of key importance for test ban monitoring. Traditionally, seismology has divided such observations into local and regional, covering the distance range up to some 2000 km, and teleseismic, covering the distances 2000–10 000 km. Beyond 10 000 km there is a shadow zone for P waves due to the earth's core, although signals are still observed beyond this range, as seen in Fig. 2.6. A seismological station can thus effectively monitor events up to a distance of 10 000 km, corresponding to half the earth's surface (Fig. 2.7). The transmission properties through the interior of the earth are fairly homogeneous and standardized global travel time models have traditionally been used for event location. P waves propagate with high velocity, reaching 2000 km in about 4 min and 10 000 km in 13 min. Recent results, including those from nuclear explosions, however, show significant variations in transmission properties around the world, giving rise to location bias and strong variation in the appearance of the signals, also for observations at teleseismic distances. Calibration is thus needed to obtain event locations with high precision using teleseismic observations.

Calibration is a necessity for observations at local and regional distances. At such distances signals propagate through the surface layers of the earth, referred to as the crust and upper mantle, where the propagation velocities and other transmission properties vary most significantly from one region to another. Over the last few decades extensive calibration work has been carried out in most regions of the earth and in most areas the location bias is now substantially reduced. Calibration work has also been carried out as part of the

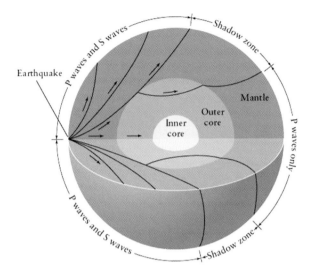

Fig. 2.7 Cross-section of the earth with the paths of seismic rays. The mantle is solid, the outer core is fluid and the inner core is again solid

build-up of the CTBT verification system. Local and regional observations are today an important element of seismic monitoring. Especially in continental areas, monitoring at distances up to 2000 km might detect events one or two orders of magnitude smaller than those that can be detected by teleseismic observations.

2.3.3 Seismological Stations and Networks

As early as 132 BC the first seismoscope, showing the direction of incoming earthquake waves, was developed in China (Inventors 2007). 2000 years later, in the late 19th century, the first seismometers were developed and installed at observatories around the world (Dewey and Byerly 2007). Seismological stations have benefited from 100 years of technical and scientific development, going from smoked paper and photographic recordings to highly sensitive, fully automatic, digital stations. The frequency band of the instruments used for global monitoring has traditionally been in the range 0.5–5 Hz. Modern instrumentation permits the use of a considerably broader frequency band. Higher frequencies are important in monitoring at local and regional distances. Modern instrumentation is capable of resolving and adequately representing ground motion of the order of 1 nm at frequencies of interest to seismological monitoring. Such small signals can be detected at low-noise sites. Advanced array stations, some with hundreds of sensors (see Box 2), have enhanced the capabilities to detect and provisionally locate weak events.

A large number of seismological stations are or have been operating around the world. The US Geological Survey (USGS) maintains an International Registry of Seismograph Stations (NEIC 2007) containing more than 10 000 stations, some of which are closed today. Many of the stations are monitoring local seismicity but several extensive networks for global earthquake monitoring also exist (IRIS 2007, USGS 2007a). Today some 4000 stations report routinely to the International Seismological Centre (ISC, see below), but the number of operational stations might be considerably higher, as many of the stations monitoring only weak local seismicity may not report to the ISC. A large number of mobile stations also exist, which can be rapidly deployed to monitor aftershocks or to study the structure of the earth. Some States have also established global networks under their national control to monitor nuclear explosions. The network established by the US Air Force Technical Applications Center (AFTAC) is an example of such a global network.

National and international organizations are compiling seismological data and reports from stations either on a global, regional or national basis, primarily for hazard mitigation. The USGS (2007b) provides a fast, global service, the European-Mediterranean Seismological Centre (EMSC 2007) covers the European and Mediterranean area, and the Japan Meteorological Agency (JMA 2007) monitors Japan and adjacent areas. The International Seismological Centre (ISC 2007) is tasked with providing the most comprehensive compilation possible of all observed earthquakes. Based on reports from more than 4000 stations, the ISC compilation for 2004 comprises some 250 000 events.

2.3.4 Seismology and Nuclear Explosions

The relation between seismology and nuclear explosions has two faces. Nuclear explosions have provided seismology with well-defined, simple sources generating signals that can propagate around the globe, and seismology has provided a key tool for monitoring nuclear explosions. A very early, perhaps even the first contribution in this regard was Beno Gutenberg's paper (Gutenberg 1946) on an interpretation of records of seismic (and also infrasound) waves from the "Trinity" nuclear test on July 16, 1945.

Nuclear explosions have provided well-controlled, precisely located and symmetric sources at many places worldwide. Very few such sources existed prior to the underground testing of nuclear explosions. A nuclear explosion is by and large a symmetric source, generating very similar seismic signals in all directions. It was a surprise to seismologists to find that the observed signals differed greatly from station to station, both in strength and frequency. Similar differences had earlier been observed from earthquakes but were attributed primarily to non-symmetric sources, radiating different signals in different directions. The fact that observations from explosions also exhibit large differences made experts realize that the earth is more complex than they had originally assumed. This most important insight promoted extensive work on the study and mapping of the structures and the wave propagation properties globally and on regional and local scales. One important finding was bias in earlier seismic travel time models. The explosions contributed to the improvement of these models both globally and on a regional scale, and seismic travel times are by now fairly well understood in most parts of the globe. Less well understood is the way the seismic signals are affected in terms of signal strength, waveform and spectral content. These dynamic parameters carry information on the source and the earth, which we have still to understand fully (Fig. 2.8).

Since the early 1960s seismology has developed a special branch to study the monitoring of nuclear explosions, usually referred to as Detection Seismology. Over the years, national programs have allocated large amounts of money to scientific studies of how to detect, locate and identify nuclear explosions

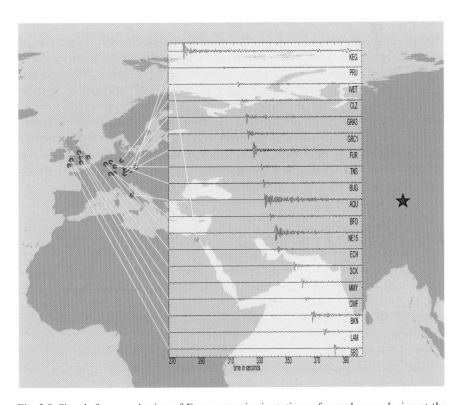

Fig. 2.8 Signals from a selection of European seismic stations of a nuclear explosion at the Lop Nor Test site in China on 21 May 1992, at 04:59 GMT. The estimated yield of this test was 660 kt. The differences in the waveforms are mainly caused by differences in the geology of the wave propagation paths to the stations

(Office of Technology Assessment US Congress 1988). Advanced station concepts, including array stations and ocean bottom installations, were developed and tested, and a large number of stations were established. The first global network of standardized stations, the World Wide Standard Station Network (WWSSN) with 120 stations in more than 60 countries and islands, was established as part of the US monitoring efforts within the VELA Uniform program. Many other countries, including Australia, Canada, China, Finland, France, Germany, the Netherlands, Norway, the Soviet Union, Sweden, Switzerland and the UK, have provided not only a good basis for implementing the seismological part of the CTBT monitoring system but also a shared understanding of how to analyze and interpret seismological data. This early work has paid off well and is the reason why the seismological component of the CTBT monitoring system could provide test data well in advance of the other technologies.

2.4 Hydroacoustic Monitoring

Hydroacoustic monitoring relates to the detection of events or objects using acoustic signals in water. The technique is most frequently used to detect submarines. The book *"The Hunt for Red October"* (Clancy 1984) gives a good insight into this military application. The CTBT application is quite different, using observations at great distances to observe events in or close to oceans. A few institutions are conducting research using experimental stations or data from military installations such as the US Navy's Sound Surveillance System (SOSUS) in the North Atlantic (NOAA 2007a). The hydroacoustic component of the CTBT monitoring system is thus a unique global system (Hall 2007, Lawrence and Galindo Arranz 2007, Jans 2005).

2.4.1 The SOFAR Channel

Low frequency sound can travel for very large distances through the oceans thanks to the SOFAR sound channel. The depth of the SOFAR channel varies in different oceans depending on salinity, temperature and the depth of the ocean. The normal range is 600–1200 m, deepest in subtropic regions and more shallow at higher latitudes (NOAA 2007b).

The SOFAR channel is a low velocity channel where sound waves get trapped and can propagate without significant loss of energy (Fig. 2.9). The propagation is so efficient that explosion charges of the order of 1 kg detonated in the channel can be detected at a distance of 10 000 km (Bannister et al. 1993). Even if stronger sources are needed for events occurring outside the SOFAR channel, as only a small fraction of the generated energy finds its way into the channel, hydroacoustic monitoring is most efficient.

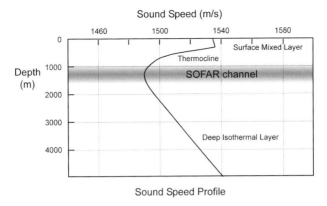

Fig. 2.9 Hydroacoustic velocity profile which shapes the wave propagation in the oceans. Sound waves propagate efficiently to very long ranges in the SOFAR low-velocity channel

2.4.2 Detecting the Signals

The most efficient way of detecting hydroacoustic signals is to place the hydrophone sensor in the SOFAR channel. To detect signals from large distances hydrophones operate at frequencies of 1–100 Hz. Normally a station comprises a few sensors to form an array that will give the direction of the incoming signal.

To place a sensor at a depth of about 1 km might present logistical challenges. There are two principal approaches: to support the sensor from a buoy, for wireless data transmission, or to connect the hydrophone to a shore station by an underwater cable. All six hydrophone stations in the monitoring network of the CTBT are cable mounted, which is quite an expensive undertaking.

An alternative, cheaper way of monitoring hydroacoustic signals is to use a seismometer placed onshore close to a steeply shelving coast. Such a station will be able to observe a seismic wave generated by the conversion of the hydroacoustic signal at the ocean-land boundary. Such converted waves are referred to as T-phases. Seismic stations close to steep shorelines normally record T-phases from events in or close to the oceans. The hydroacoustic network of the CTBT monitoring system includes five seismic stations placed on steeply sloped islands. These stations are specially tuned to record T-phases and operate in a frequency band of 0.5–20 Hz, a little higher than normal seismic stations.

2.4.3 Locating Events

Thanks to the high detection capability provided by the SOFAR channel, oceans can be confidently monitored by a small number of stations. What about the location accuracy? Hydrophones operate at higher frequencies than seismic stations, and this normally improves the accuracy with which the arrival time of a signal can be determined. The propagation speed is low (about 1.485 km/s) compared to seismic signals and this reduces the consequences of timing errors. Hydroacoustic networks should thus in principle be able to locate events with high accuracy. The small number of stations (only 11 in the IMS), however, limit the accuracy of event location using hydroacoustic information only. No data are currently available on the location uncertainties, but they are likely to be larger than for the seismic network. Quite often, seismological and hydroacoustic data can be used together, which reduces the location uncertainty.

2.4.4 What do We Observe?

In addition to observing explosions in the oceans, what else is detected? Hydroacoustic signals, mostly in the form of T-phases, are observed from earthquakes that occur under the oceans or in the direct vicinity of shorelines. Underwater volcanoes are another source of hydroacoustic signals. Man-made activities, such as underwater construction work and seismic exploration surveys, are also observed. Interestingly, from an environmental perspective, the calving of icebergs can be detected by hydrophones.

2.5 Infrasound Monitoring

Infrasound monitoring is the youngest branch of the four monitoring technologies (Gossard and Hooke 1975, Le Pichon et al. 2009). It is used to monitor atmospheric explosions and complements radionuclide monitoring. The main reason for having both technologies is that infrasound measurements provide rapid event locations, while radionuclide measurement is decisive with respect to the identification of nuclear events.

During the negotiations of the CTBT only a handful of States – Australia, France, the Netherlands, Sweden, UK and the USA – had experience with the use of infrasound as a detection technique. Moreover, the existing infrastructure was very limited, in contrast to seismology. The entire network of 60 infrasound stations that is part of the CTBT monitoring system thus had to be designed and built from scratch.

2.5.1 What Is Infrasound?

Infrasound is just normal sound, but of very low frequency. The name is similar to the term infrared used in the optical spectrum, referring to the part of the spectrum having lower frequencies than red light. The infrasound spectrum ranges from the sub-audible sounds at 20 cycles per second, just where the sub-woofer of an audio system will produce only very limited output, to the very low frequencies when one cycle lasts for 100 s or more.

A remarkable property of infrasound waves is their low attenuation or damping in the atmosphere. Due to the composition of the atmosphere, which consists mainly of nitrogen and oxygen, the absorption of infrasound is very low. The lowest frequencies have the lowest absorption. Water vapour, which is present in low concentrations in the lower atmosphere, will increase the absorption somewhat in the lowest part of the atmosphere compared to a dry atmosphere. Sound waves have almost the same velocity of propagation, around 300 m/s, independent of the signal frequencies. The temperature variation with altitude contributes to the long-range sound propagation properties in the atmosphere. The temperature structure, and as a consequence the velocity structure of the atmosphere is such that the sound waves are generally reflected at two altitudes, one at 30–50 km and one at 100–120 km. These reflectors bend the signals back to the earth's surface, where they again bounce back into the atmosphere. In this way, through a series of reflections, infrasound signals can propagate over large distances of thousands of kilometres.

Unlike the circumstances governing the propagation of seismic and hydro-acoustic waves, the medium for infrasound propagation is a dynamic one, changing its properties over time. This is caused mainly by winds, which change both in time and space. At high altitudes strong winds of 50 m/s are common, and winds above 120 m/s are sometimes observed. These wind speeds are substantial fractions of the velocity of sound. The propagation properties of the infrasound signals are thus strongly influenced by strong winds and the geometrical spreading of the sound waves they cause.

2.5.2 Detecting Signals

Infrasound signals are detected by highly sensitive barometers which measure atmospheric pressure. These microbarometers, as they are called, are attached to devices designed to reduce the pressure fluctuations from the ever-present wind and turbulence in the atmosphere. These noise reducing pipes, having a number of discrete inlets, are connected to the microbarometer. The pressure that is presented to the microbarometer is proportional to the summed inputs of the discrete inlets. In this way the pressure field is averaged over an area, and atmospheric disturbances of short wavelength are averaged out. The inlets are at the surface of the earth, where wind speeds are the lowest due to friction with the ground. Reduction of the noise by a factor 10 or more can be achieved in this way. Noise in infrasound measurements is further suppressed by combining 4–8 microbarometers in an array configuration, just as in seismology or hydro-acoustics. The array configuration for stations used in CTBT monitoring usually has an aperture, or overall diameter, between 1 and 3 km. Over these distances infrasound waves are generally coherent and of the same shape, which is a necessary condition for archiving a good gain in the signal-to-noise ratio.

In order to gain an impression of the signal strength, the following example may be illustrative. We use as a measure the pressure gradient that is present in the earth's atmosphere. At sea level the pressure is about 100 000 Pascal, which equals one atmosphere. The pressure decreases with altitude and the pressure difference, or gradient, over 10 cm is typically 1 Pascal. This corresponds to a sizable infrasound signal. Usually signals as low as 0.01 Pascal can be detected, which corresponds to the gradient over just 1 mm. So if we, as an experiment, lift the microbarometer just 1 mm, this would simulate the amplitude of a typical infrasound recording.

In designing infrasound arrays for the CTBT monitoring system, great care is taken to find a cost-efficient design, adapting the mechanical layout, such as the number of discrete inlet ports and number of array elements, to the local conditions. For instance, in a tropical rainforest the conditions for infrasound detection are quite favorable, as the dense forest acts as a natural noise reducer. Windy island sites in the middle of oceans and rough open landscapes with strong turbulence, on the other hand, are not optimal for detecting infrasound signals. An extensive program of site surveys has been necessary to find good locations at, or close to, the locations identified in the Treaty. The limited experience so far with infrasound station observations indicates that low level

Fig. 2.10 Not all mushroom clouds are of nuclear origin. The picture shows an eruption on 7 October 1999 of the Guaqua Pichincha volcano near Quito, Ecuador. Detection of infrasound waves from volcanic eruptions can be used to warn air traffic of *ash clouds*. Photo courtesy of Patty Zway and the Geophysical Institute, Escuela Politecnica Nacional, Quito

signal detection can be achieved, given that stations are carefully designed and implemented. The joint analysis of data from many infrasound stations, and optimizing the capability of the infrasound network, remains a challenge.

It has gradually become clear that the sources of infrasound are more plentiful than was originally expected. Sources of different kinds are known by now, and are still being studied. They range from meteorites and volcanoes (Fig. 2.10) to interfering waves in the ocean basins and lee waves behind mountain ranges. Infrasound signals have also been observed from earthquakes and from the aurora borealis. The location of infrasound sources uses similar methods as in seismology and hydroacoustics.

The introduction of infrasound, as one of the three waveform monitoring technologies in the CTBT, provided a big boost to research in this field, which lacked almost any interest from the time nuclear testing moved underground more than 25 years ago. In the last ten years, a new and stimulating branch of atmospheric science has developed with connections to seismology, hydroacoustics, and meteorology.

2.6 Radionuclide Monitoring

To detect radionuclide material, be it particles or noble gases, that can be traced back to a nuclear explosion is indeed to find the smoke from a smoking gun. Radionuclide detection methods are thus a most important component of a CTBT verification regime, both as part of the global monitoring system and as a tool in on-site inspections.

2.6.1 Particles and Noble Gases

A nuclear explosion produces large quantities of radioactive fission products. Their composition depends to some extent on the fissile material, uranium or plutonium, used in the nuclear device. The initial nuclear radiation from an explosion, especially neutrons from thermonuclear devices, will also interact with the surrounding materials, in the device and in the environment, and create radioactive activation products. Many of the radioactive products decay and disappear rapidly, while others have half-lives that well fit the CTBT verification scheme; long-lived enough to survive the several days time scale of sampling and measurement but also short-lived enough to emit substantial radiation and not to stay around too long to create a "natural" background. Among the most potent nuclides that signal a nuclear explosion are zirconium-95 and -97, molybdenum-99, iodine-131, -132 and -133, ruthenium-103 and barium/lanthanum-140 (De Geer 2001). A nuclear explosion also creates radioactive noble gases, which play a special role in detecting and attributing underground nuclear explosions. Due to their chemical "nobility" they do not adhere to surfaces underground and therefore they can readily leak out of cracks and fissures and be detected in the atmosphere days, weeks and months after an explosion. The CTBT verification system focuses on a quartet of xenon isotopes and metastable states (m) (xenon-131m, -133m, -133 and -135) which have suitable half-lives and radiation for detection.

In an atmospheric explosion or an explosion at the surface of the earth the radioactive products are ejected as plumes high into the atmosphere. The debris can then be carried by winds and air circulation for thousands of kilometers. In an underground explosion most of the fission products are contained underground in the cavity and its close vicinity and well contained tests should not leak radioactive particles into the atmosphere at all. There are, however, examples when fission products have vented into the atmosphere and have been detected at large distances (Layton 2007, De Geer 1991). As already alluded to, radioactive noble gases are more difficult to contain and xenon has been observed from several underground tests in the past. The latest indication is the xenon-133 observation at the IMS station in Yellowknife in Canada, which was consistent with a release by the North Korean explosion on October 9, 2006 (Saey et al. 2007). Xenon isotopes were also detected by a Swedish mobile system in the very north-east corner of South Korea a few days after this test (Ringbom et al. 2007).

2.6.2 Particle Detection at CTBT Monitoring Stations

The equipment for collecting particles is in principle very simple; a large fan that blows air through a filter. Large volumes of air, of the order of 1000 cubic m/h, flows through a filter paper that collect particles with a high efficiency. The filter paper is exposed to the air flow for 24 h. It is then compressed and 24 h later its radioactivity is measured by a high-resolution gamma detector for 24 h. Here too we have background noise, much of it coming from natural radon in the atmosphere. Parts of this radiation are short-lived and this part is eliminated by delaying the measurements of the filter by one day. The measurement of the filter gives a gamma-ray spectrum (Fig. 2.11), in which the intensity of the radiation is plotted for all different energies. The content of the sample can be analyzed from manifest peaks in the spectrum that correspond to different radioactive isotopes. A sample or a spectrum is categorized into one of five levels, depending on which radionuclides have been detected. For this purpose 84 radionuclides have been defined as relevant for screening, 42 being fission products and 42 being activation products (De Geer 2001). A Level 1 is a spectrum containing normal amounts of natural radionuclides. A Level 5 contains at least two of the 84 defined relevant nuclides and at least one of them should be a fission product. This scheme is meant to help States focus their interest on relevant spectra. A Level 5 sample is also subject to further analysis at two radionuclide laboratories to confirm the measurement and, with better equipment and a lower noise level, possibly also to detect other relevant nuclides.

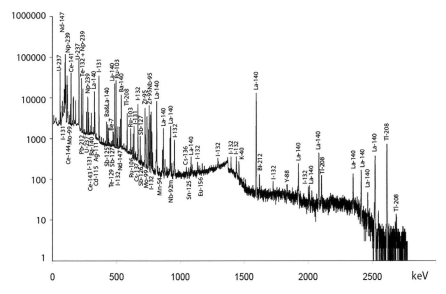

Fig. 2.11. Gamma ray spectrum from nuclear fallout. The important isotopes are marked. The *horizontal axis* represents the energy of gamma rays in keV, which is the equivalent energy of an electron when it is accelerated in an electric field of 1000 Volts. *Vertical axis* represents the normalized number of measured counts

2.6.3   Detection of Noble Gases at CTBT Monitoring Stations

The equipment needed to collect and detect radioactive noble gases is more complex than that needed for particle detection. To be efficient for monitoring purposes a noble gas detector must, in a way similar to the particle detector, concentrate the radioactive gas before any measurements can be made. This concentration is a more difficult process for noble gases. Currently xenon analyzers developed by three CTBT States Signatories – France, the Russian Federation and Sweden – are deployed and they utilize somewhat different techniques to concentrate the xenon gas. They all use chromatographic separation on activated charcoal, the French with a special added membrane that enhances the separation, the Russian system processing at quite low temperatures (down to –170°C), and the Swedish processing at room temperature. Test measurements have commenced at some of the stations in the CTBT monitoring system and these are discussed in later chapters.

The detection of radioactive noble gases also suffers from a background noise problem due to natural radiation, which may mask the radiation from xenon. Much of that, however, can be reduced by an ingenious use of coincidence techniques where two simultaneously emitted radiations typical for the xenon isotopes are required for detection. Radioactive xenon isotopes can also originate from other sources, most notably nuclear power plants and facilities for producing isotopes for medical purposes. The extent and origin of such sources need to be further studied, and in addition to the noble gas stations being tested as part of the establishment of the CTBT monitoring system, the European Union is financing a project to increase our knowledge at a number of sites around the globe.

2.6.4 Tracking the Source

Radionuclide measurements are able to detect low concentrations of radio-active particles or gases and to provide comprehensive information to facilitate the identification of the source. The Achilles' heel is the location uncertainty. Radionuclide gases or particles are transported from the source to observing stations by the movements of air masses. To locate the source from observations you have to retrace that journey. The radionuclide measurements might give some information on when the products were formed and thus on how long time they have been travelling. To calculate the track the air volume containing the radioactive material has travelled before it arrives at an observing station requires the use of a complex atmospheric transport model and comprehensive meteorological data (Becker et al. 2007, Wotawa et al. 2003, Issartel and Baverel 2003). Even using all the data available to large international weather centers, it is inherently impossible to define an exact track along which the source is to be found. Due to diffusion processes only an area can be defined where a certain size emission is consistent with the observations. If several samples show detections, these areas will naturally reduce.

In general, however, there is substantial uncertainty in locating sources using radionuclide data alone, which can be measured in hundreds of kilometers or more. Given this great uncertainty, and also the fact that radionuclide measurements are the only way to determine that an event is nuclear, it is essential to benefit from synergies between the different components of the verification systems. Seismological and hydroacoustic observations can provide a quite precise location and time of origin of an event, whereas infrasound observations give greater uncertainties. The xenon observation, mentioned above, at the Yellowknife station in Canada, consistent with a possible release from the North Korean explosion on October 9, 2006, is an interesting example of this process (Fig. 2.12). This is a most obvious example of synergy between different components of the CTBT verification regime.

2.7 Detection from Space

In 1963, three days after the Partial Test Ban Treaty was signed, the first pair of Vela Hotel monitoring satellites were launched by the USA (Decode Systems 2007). They were placed in orbits at an altitude of about 100 000 km. In all 12 Vela satellites were launched, the last one in 1970, and the last satellite was in operation until 1984. The initial six satellites were equipped to detect X-rays

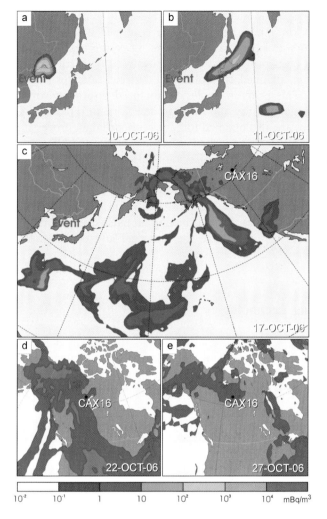

Fig. 2.12 Tracking of the xenon-133 isotope from the North Korean test explosion on October 9, 2006 to Yellowknife (CAX16) in Canada. *Color coding* indicates the expected concentration of radionuclide activity

and neutron and gamma radiation from nuclear explosions in the atmosphere and in space. The last six Advanced Vela satellites also had systems, so-called "bhangmeters", to detect the double pulsed light-flash characteristic of a nuclear explosion and also systems to detect electromagnetic pulses. It was one of these Vela satellites that detected such a light flash over the South Atlantic Ocean on 22 September 1979. The origin of this observation is still unresolved (NSA 2006).

The Vela system has been replaced by the US Integrated Operational Nuclear Detection System. This is based on sensors placed in 2001 on 33

Navstar Global Positioning System satellites in considerably lower orbits of about 20 000 km. These satellites were equipped with optical, neutron and X-ray detectors and gamma-ray sensors. Over the next decade an extensive program has been proposed for the sensors to be carried by the follow-on satellite system, planned to replace the current satellites beginning in 2006 (NAS 2002). This program includes enhanced optical sensors, autonomous EMP sensors, gamma-ray and neutron detectors, infrared and X-ray sensors. These satellites are also planned to have on-board data processing capability.

The satellite systems described above are part of the US national technical means and not of the international verification regime. It is interesting to note that the US is planning a considerable modernization and upgrading of these systems, even at a time when it is unlikely that any nuclear test will be conducted in the atmosphere. We have no information on whether similar systems are operated by other states. Russia, having an extensive and multifaceted space program, would certainly have the technical and operational capability to establish and operate such a system.

2.8 Enabling Technologies

The development and establishment of the verification regime has been greatly facilitated by the scientific and technological developments, not only within each of the verification technologies mentioned above but also within other more generic areas such as computer hardware and software and communications. These developments have to a large extent been incorporated in the verification system and have improved it substantially. For a variety reasons, other developments have not been incorporated, such as satellite based observation and navigation systems, which, for political reasons, were rejected during the negotiations. We here briefly discuss these enabling technologies, both those that have been used and those that have not.

2.8.1 Computer Hardware and Software

The dramatic development over the last two to three decades in computer hardware, dramatically increasing computing speed and memory while at the same time reducing cost, has in fact been a prerequisite for the present CTBT monitoring system. This development, following what has been called Moore's law, is illustrated in Fig. 2.13. In the early days of the development of a global verification system the lack of computing power and storage memory was a limiting factor, even with access to some of the most powerful computers available on the market at that time. Today speed and memory are no longer limiting factors for CTBT verification applications. New software development tools have also proved to be most important. Information handling and

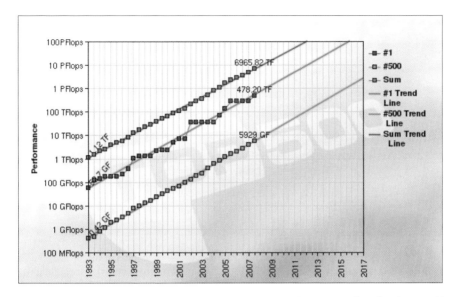

Fig. 2.13 Development of computer speed over the last decade. Data used are for the top 500 computers (Top500 2007)

graphical presentation programs have become available and new languages have made program development more efficient.

2.8.2 Communications

Global communications have moved from telex to broadband. When the international cooperative work started in the mid 1970s only very limited amounts of information extracted from the recordings could be exchanged. Today all waveform data are collected on-line and it has even proved easier and less costly to collect all data, even from stations that should be reporting short segments of data on demand only. The development of global communications has thus been as important to the verification regime as that of the computers. A specially designed communication system for CTBT verification purposes, based on modern satellite technology, the Global Communications Infrastructure, has been established to connect monitoring stations and National Data Centers around the world with the International Data Center in Vienna.

2.8.3 Satellite Observations

For political reasons satellite observations were not included in the CTBT verification regime. "Bhangmeters" were considered too much of a national technical means to be shared among all States. In the mid-1990s high-resolution

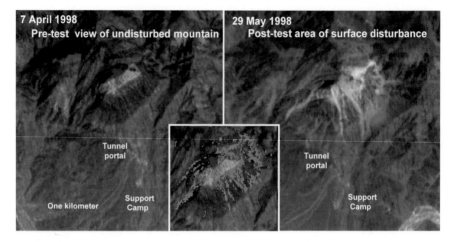

Fig. 2.14 Satellite photos from a test site in Pakistan showing changes at the surface caused by the nuclear test explosion on 28 May, 1998. In the insert, changes detected are highlighted in *red*

satellite observations were not at anyone's fingertips. High-resolution images were still considered a national asset and images that were available were difficult and costly to obtain. Google Earth and other tools to reach and present satellite photos from any part of the world with high resolution and without any cost have totally changed the picture. Readily available satellite images with high resolution would no doubt be most useful in the analysis and interpretation of observed events and in the eventual preparation for an on-site inspection. Such images would be especially powerful if a scene can be analyzed before and after an event has occurred to see what may have changed (Fig. 2.14).

2.8.4 Revolution in Information Analysis

The last decade has seen a most dramatic development in the area of data analysis and data management, usually referred to as data mining. Advanced database techniques have paved the way for new approaches to searching, combining and analyzing large amounts of information. Different web applications have prompted this most rapid development. The core analysis components of the extensive computer program package used to analyse data, in particular the seismic data, from the CTBT monitoring network were to a large extent developed in the late 1980s or early 1990s and do not, for obvious reasons, contain any of those new developments. We might expect that the use of new data mining techniques would substantially improve the quality of automatic data analysis and provide more information on the events observed.

Box 2
Array Stations

An array station is a kind of antenna, composed of a number of sensors placed in a geometric pattern, to enhance the ability to detect weak signals and to provide a preliminary location of the event. The photo in Fig. B2.1 is an example from an infrasound array, where atmospheric pressure is summed over the inlets to improve detection of infrasound signals by enhancing the signal-to-noise ratio (SNR). The concept of array stations was first developed to detect radar signals and the first seismic array designs were developed by radar experts. Based on experience from radar stations, a Large Aperture Seismic Array (LASA) (Green et al. 1965) was built in Montana, USA, in the 1960s, with more than 500 seismometers spread over an area with a diameter of 200 km. This turned out to be a successful failure. It was a failure in the sense that it did not provide the expected gain as the seismic signals did not behave like the radar signals but changed most significantly over those distances. It was successful in the sense that it provided most useful information on the fine structure of signal propagation and the characteristics of seismic noise, of great value for the design of the next generations of array stations.

The basic idea behind an array is that the signals appearing at the different elements of an array should be identical and only shifted in time, whereas the noise should be different from element to element. In other words, the signals should be coherent and the noise incoherent (Fig. B2.2). In what is referred to as "beam-forming", signals from the individual sensors are added together after introducing time shifts to bring signals with a particular direction and velocity in phase. If the signals are fully coherent, the added signal will increase in strength roughly proportionally to the number of sensors (N). The noise will also be added, and if the noise is fully incoherent, meaning that there is no propagation of noise from one sensor to the next, the strength of the noise will increase with the square root of N. The SNR would thus increase by the square root of N. For a 25 element array this implies an increase by a factor of 5 of the SNR. As these assumptions, fully coherent signals and non-coherent noise, are not fully satisfied in real life, the gain is normally a little less than the square root of N.

Fig. B2.1 IMS infrasound station IS18 at Qaanaaq, Greenland, showing the array concept. Here atmospheric pressure is averaged over the inlets located at the end of the pipes laid out in a star pattern

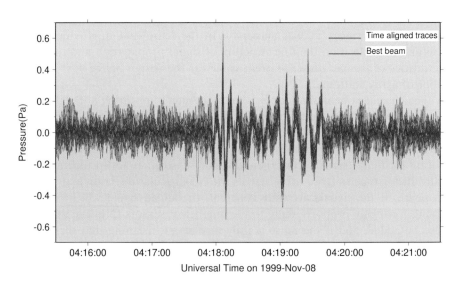

Fig. B2.2 A signal propagating across an array reaches individual elements at different times. Array processing involves the time shifting of individual traces (in *blue*) to achieve alignment. When these traces are summed the signal (in *red*) becomes clearly visible

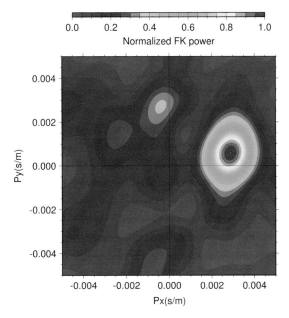

Fig. B2.3 Frequency-slowness power density spectrum for the signal of Fig. B2.2. This spectrum can be seen as the result of the systematic shifting and summing of the individual traces of the array. For any point of the plot a velocity and a direction of the signal is assumed and the resulting (normalized) power of the summed traces is plotted. The axes show the "slowness" (the reciprocal velocity). The peak of the plot (marked in *red*) corresponds to the velocity and direction of approach of the signal. The plotted spectrum can also be interpreted as an "all sky camera image" of the source

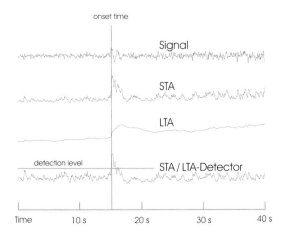

Fig. B2.4 Illustration of the detection process. The *top trace* is a seismic recording and the STA trace is the short term average which is sensitive to arriving signals. The LTA trace is the long term average, which provides information on the noise level at the site. The *bottom trace* represents the ratio STA/LTA, which is sensitive to signals and at the same time adjusts the sensitivity to the actual noise level. A level is set to declare a detection of a signal

In the processing of array data a large number of "beams", normally several hundred, are formed, corresponding to different directions and apparent velocities. Computing a large number of beams in real time is no challenge to modern computers. The apparent seismic velocities for signals of interest range from about 6–24 km/s, the velocity being higher the further away the source is. Each beam thus looks at a certain area of the globe. Apparent velocities and directions of approach are estimated using so-called frequency-slowness analysis techniques, as illustrated in Fig. B2.3.

To detect signals each beam is filtered in the frequency band that gives the highest SNR for signals on that particular beam. Optimization requires different filtering for signals from events in different regions. The detection algorithm then compares the short-term variation of the data – a short-term average over a few seconds – to the long-term variation – a long-term average over about a minute. In this way the detector "floats" on the noise and adjusts the detection capability to the actual noise, giving higher capability when noise is low and vice versa (Fig. B2.4). A detection on a certain beam also indicates a rough location of the event.

Chapter 3
A Long Journey to a Treaty

Test ban negotiations proved to be a 50-year journey from the first initiative until the CTBT was signed. Still, after more than another ten years, we have not yet reached the final destination; a CTBT that has entered into force. We may have learned two things from all these years. International negotiations take time, move in small steps and require patience and further, that the possibility of making progress is held hostage by the general political situation, especially among the key players.

Over the years a large number of initiatives have been taken, some bilateral between the USA and the USSR, while others were trilateral, also including the UK, and yet others were multilateral. These initiatives kept the issue alive and the pot, while not boiling, was at least lukewarm on the back burner and not in the deep freeze. The results were a number of partial steps forward, some of which were reversed. Unilateral or bilateral moratoria and agreements on partial measures, such as banning explosions in the atmosphere and banning underground tests above 150 kt, were such important steps. This long period clearly showed the importance of not giving up or becoming dispirited, but of continuing to try, accepting partial measures, imperfect as they might be. A step-by-step process might stand a better chance of leading to the desired goal than an all-or-nothing approach. A step-by-step process might by itself build the mutual trust and confidence necessary for a final agreement.

3.1 Past Test Ban Initiatives

The question of how to ban nuclear explosions is almost as old as the testing itself. The intent here is to briefly summarize the different initiatives that were taken before the final CTBT negotiations started in 1994. This summary is based on material presented by Ramaker et al. (2003) and Dahlman and Israelsson (1977). A number of other books reflect different perspectives on the CTBT negotiations and the CTBT (Hansen 2006, Goldblat 2002, Goldblat 2000, Goldblat 1988, Avenhaus et al. 2002).

O. Dahlman et al., *Nuclear Test Ban*, DOI 10.1007/978-1-4020-6885-0_3,
© Springer Science+Business Media B.V. 2009

3.1.1 First Efforts

As early as June 1946 the USA presented the Baruch plan (Atomic Archive 2007a), named after Bernard Baruch, a businessman turned presidential advisor. The plan proposed nuclear disarmament and the establishment of an International Atomic Development Authority using international inspectors to control all production of fissile material. The USSR did not like the idea of inspection of all fissile material before disarmament. Their counter-proposal was the Gromyko Plan, published in the same month, which proposed to start with the dismantling of nuclear weapons, meaning the US weapons as the only ones existing at that time, and then having less stringent inspections. The Gromyko plan was rejected by the USA and the bilateral discussion of the issue came to an end.

On April 2, 1954 Prime Minister Nehru (Nehru 1954) of India took a first international initiative by calling for a "Standstill agreement" on nuclear testing. This initiative followed a US high-yield thermonuclear test explosion in the Pacific one month earlier, which had caused global concern as widespread fallout irradiated people living in the area and the crew of the Lucky Dragon fishing boat. Mr Nehru proposed that discussions on such an agreement should begin in the UN Disarmament Commission, which was established in 1952. Nehru also proposed that information on the effects of nuclear weapons should be released by the nuclear weapons States, to create a public pressure against such weapons.

3.1.2 The Five

During the second half of the 1950s a number of arms control proposals related to a test ban were made and rejected. The Sub-Committee of Five – Canada, France, the UK, the USA and the USSR – was formed within the Disarmament Commission and at a meeting of that sub-committee in May 1955 the USSR proposed the discontinuation of nuclear testing, to be monitored by an international commission. After a year of discussion the proposal was finally rejected. Also in 1957, proposals were made by both the USA and the USSR to halt testing for a period of two to three years. The terms attached to the proposals proved unacceptable to the other party, however.

In March 1958 the USSR announced a unilateral moratorium on nuclear testing as long as other nuclear weapon States refrained from testing. In July 1958 the first international meeting on nuclear test verification was held at the UN in Geneva and was attended by scientific experts from eight States: Canada, Czechoslovakia, France, Poland, Romania, the UK, the USA and the USSR. Their report (Strickland 1964) contains a proposal for the establishment of a network of 160–170 control posts on land and 10 on ships. This system was expected to detect, locate and identify atmospheric and underwater explosions with yields above 1 kt and underground explosions with yields above 5 kt. In fact the proposed system was very far removed from the global monitoring network of the CTBT, negotiated almost 40 years later, involving some 300 stations, although the technologies and capabilities were different. The concept of on-site inspections was also introduced. Whether such inspections should be voluntary or mandatory was an issue of contention for a long time.

3.1.3 The Trilaterals

In October 1958 the UK, the USA and the USSR commenced trilateral test ban negotiations and from the beginning of 1959 the three States observed a moratorium on nuclear testing. In the USA the Berkner Panel (Barth 2003) found that the system developed during the Geneva expert meeting in 1958 would not offer the expected capability, and further international experts' meetings were held in the second half of 1959. The Panel also laid out the strategy for an extensive US research effort on test ban verification, known as the Vela project. Richards and Zavales (1996) give a good account of the technical issues during this period. In February 1960 the USA tabled a proposal on a treaty banning tests in the atmosphere, under water and those underground tests that would create a seismic magnitude greater than 4.75, corresponding to a yield of about 5 kt. In response the USSR proposed that the ban should include nuclear tests in space and a four to five year moratorium on underground tests below the 4.75 threshold. It was expected that the summit meeting between USA and the USSR in Paris in mid-May 1960 would resolve a number of outstanding issues that were impeding progress towards a treaty. The meeting was dominated, however, by the shooting down of the American U2 aircraft over the USSR on May 1, 1960.

When the trilateral test ban discussions were reconvened in March 1961, the UK and the USA tabled a draft treaty banning all tests except underground tests that generate a seismic magnitude less than 4.75, linked to a three-year moratorium on all tests. The draft treaty contained provisions for a quota system for on-site inspections, allowing 20 inspections per year in the USSR and a corresponding number in the USA. The USSR offered three inspections per year and the discussions got bogged down.

Dramatic world events took over. The first one was the US attempt to invade Cuba at the Bay of Pigs in April 1961. In mid-1961 too the crisis over Berlin escalated and on August 13 the Berlin Wall started to go up. On September 1, 1961 the USSR resumed nuclear testing and the USA followed suit on September 15. The deteriorating relation between the two powers led to the most extensive testing ever observed with an average of more than two tests a week. This included the "doomsday" explosion on 30 October, 1961 at Novaya Zemlya in the Barents Sea. This is the most powerful nuclear explosion ever conducted, with a yield of about 60 megatons or some 4000 times more powerful than the nuclear weapon used on Hiroshima at the end of World War II. The radioactive fallout from the explosion created problems in the Scandinavian countries and increased environmental concern about atmospheric testing. World tension reached another serious peak with the Cuban Missile Crisis (Goldman and Stein 1997) in October 1962, with a USA–USSR confrontation on Soviet nuclear weapons in Cuba. The trilateral negotiations stalled and were eventually adjourned indefinitely.

3.1.4 The Partial Test Ban Treaty

In early 1962 the focus had moved away from the trilateral talks to the newly established United Nations Eighteen Nation Committee on Disarmament (ENDC) in Geneva. A draft text was presented for a comprehensive treaty banning all nuclear explosions. The proposed verification system included a seismic system and an annual number of on-site inspections. The USSR introduced the notion of "black boxes", or unmanned seismic stations. Arguments over the number of on-site inspections continued and came as close as six to seven in the US proposal and two to three on the Soviet side. Here the negotiations on a comprehensive ban broke down.

Following a joint UK-USA proposal the negotiations in the summer of 1963 shifted gear to aim for a partial treaty banning explosions in space, in the atmosphere and under water. After a few weeks of negotiations in Moscow between the USSR, the UK and the USA, a treaty was ready (Kimball and Boese 2003). The Partial (or Limited) Test Ban Treaty (PTBT) (CNS 2007d) was signed by UK, USA and USSR in Moscow on 5 August 1963. China and France did not sign. The PTBT could be considered a political achievement in the wake of the Cuban missile crises a year earlier. It was also environmentally important to free the world from radioactive fallout, which was of great concern in the early 1960s. As we have seen in Chapter 1, the PTBT made States abandon their very high yield testing, but it did not constrain testing in any other way. Nuclear testing just went underground and continued at a high rate.

Following a UN resolution that acknowledged the PTBT and asked the ENDC to commence work towards a comprehensive test ban, CTBT was put on the agenda of the ENDC in 1964, where it would stay for more than 30 years. The test ban issue was put on the back burner for the rest of the 1960s, when the focus was on negotiating the Non-Proliferation Treaty (NPT), which was concluded in 1968 (see below). In 1969 the ENDC was enlarged and changed its name to the Conference of the Committee on Disarmament (CCD). CCD continued the ambitious but unsuccessful elaboration of a comprehensive test ban treaty throughout the 1970s. The focus of the USA and the USSR was on bilateral negotiations that, in May 1972, resulted in the signature of the Strategic Arms Limitation Talks Interim Agreement (SALT I) (FAS salt 1 2007) and the Anti-Ballistic Missile (ABM) Treaty (FAS abm 2007).

3.1.5 The Threshold Test Ban Treaty

On the test ban issue, too, the two main powers again left the multilateral forum of the CCD and entered into bilateral negotiations. In July 1974 the USA and the USSR, after just a few months of negotiations, concluded the bilateral Threshold Test Ban Treaty (TTBT) (FAS ttbt 2007) limiting nuclear weapons tests to yields below 150 kt. The TTBT entered into force in 1990, immediately following ratification by the USA. The TTBT contained an obligation to negotiate an agreement covering so-called Peaceful Nuclear Explosions (PNEs). This agreement (PNET), which limited the yields of PNEs also to 150 kt, was signed in May 1976 but not ratified until 1990. These treaties contain detailed specifications of an extensive information exchange and they also provide for on-site observations by the other party under specified circumstances. This represented a big step forward in mutual information exchange between the two main powers.

Towards the end of 1977 the USA, the UK and the USSR took a new trilateral initiative towards a comprehensive test ban. The handling of on-site inspections remained a hurdle, as did questions related to the duration of the treaty, since the USSR wanted a treaty of limited duration, to be renewed when the other nuclear weapons States also joined. France and China had commenced testing at that time. The USA and the UK finally also agreed to a limited duration and proposed a three year treaty. The issue of PNEs was initially an obstacle but the USSR agreed to observe a moratorium on such explosions until arrangements had been made that precluded any military benefits from such explosions. Eventually, by 1979, the political momentum for a new test ban agreement dissipated due to unrelated political events including the Soviet invasion of Afghanistan and the hostage crisis at the US embassy in Iran.

In the 1980s USA and USSR adhered to the TTBT and the PNET but the USA deemed that the verification protocols did not provide the assurances needed. Over the years the USA had experienced a problem with some Russian explosions that they estimated, based on seismic data, had a yield above 150 kt. The USA based their conclusions on experience from their test site at Nevada, in a different geophysical environment. Based on experience from the Eurasian continent, some institutions in Europe found that seismic observations of the Russian explosions were consistent with an assumption that the yields were below the 150 kt limit. To settle the issue the USSR and the USA conducted a Joint Verification Experiment (Sykes and Ekstrom 1989) involving two nuclear explosions in 1988: one at the Nevada Test Site in the USA and one at the Semipalatinsk test site in Kazakhstan. Both sides were using a device placed in a borehole close to the nuclear charge to measure the speed of the initial shock-wave, giving an accurate measure of the yield. The purpose was to calibrate the test sites and thus get a reference for the yield estimates made from remote seismic observations. The Joint Verification Experiment settled the issue and was also an important confidence building measure.

3.2 Group of Scientific Experts

In addition to the political work at the ENDC and the CCD in Geneva, a number of technical experts' meetings were held. After a few such Ad Hoc experts' meetings it became obvious that this method of work would not lead to a common understanding of the design of a global verification system. Sweden therefore proposed that sustained expert work should be initiated to design and test a global verification system. On July 22, 1976 the CCD established "*an Ad Hoc Group of Government-appointed experts to consider and report on international co-operative measures to detect and identify seismic events, so as to facilitate the monitoring of a comprehensive test ban*". The Group was tasked to "*specify the characteristics of an international monitoring system*". For political reasons the Group's work was limited to seismic verification. The design in principle proposed by the Group also proved to be a good model for the other verification technologies used in the CTBT.

This group, referred to as the Group of Scientific Experts (GSE, Fig. 3.1) worked as a subsidiary body to the CCD and to the Conference on Disarmament (CD), which succeeded the CCD in 1979. The Group normally held two meetings per year, each lasting two weeks, and reported back to the CD after each meeting. A large number of experts from around the world participated in the work of GSE. GSE had two Swedish chairmen, Dr Ulf Ericsson (1976–1982) and Dr Ola Dahlman (1982–1996). Dr Frode Ringdal of Norway served as the Group's scientific secretary and contributed greatly to the Group's progress throughout the entire 20 years. Four senior political officers from the UN – Mr Csillag, Ms Waldheim, Mr Cassandra and Ms Mackby – supported the Group as secretaries over the years. The work during four weeks of meetings in Geneva every year was only the

Fig. 3.1 Photo of the Group of Scientific Experts during its session in August 1991 in Geneva

tip of an iceberg. Hundreds of people were engaged in many States to establish modern stations and data centers and to test the systems proposed by the Group.

GSE was a unique effort to promote a disarmament treaty by sustained technical work in a political environment. Twenty years is a long time and from a purely technical point of view the work could have been concluded sooner. During several years at the height of the cold war GSE was the only ongoing multilateral dialogue on disarmament issues and progress was slow. GSE also continued its activities, waiting for the CD to initiate political test ban negotiations.

The evolution of GSE work from 1976 to 1996 is an interesting journey where the course – the design in principle of the global verification system – was steady but the technologies underpinning the system developed dramatically. This was a journey in which many experts from around the world participated and contributed to the creation and sharing of knowledge on how to verify a nuclear test ban.

3.2.1 The Initial Design

Early on the Group decided on a design in principle of a global seismic verification system consisting of three main elements:

- A network of more than 50 seismological stations around the world;
- A routine, rapid international exchange of data, initially using the Global Telecommunication System (GTS) of the World Meteorological Organization (WMO);
- Processing of data at special International Data Centers (IDCs).

Data to be reported would be highly standardized and on two levels. Basic parameters of detected seismic signals determined at each station, Level I data, should be routinely reported to the IDCs with minimum delay. Observed waveform data, Level II data, should be provided in response to requests.

It is interesting to note that the initial principle design, including the size of the network, was maintained throughout the process. This was a great advantage and technologies and new methods could be added as they became available without changing the overall concept. When this first design was reported by the Group in 1978 (CCD/558 1978) very few modern seismological stations were equipped to make digital recordings. Communications were a bottleneck and the WMO GTS was the only available global communication network. This was essentially a telex network with a capability that allowed transmission of only a few printed pages from remote locations. Higher capacity networks were on the horizon, though. The American ARPANET, the forerunner to Internet for inter-computer communications, was early discussed as a model for the rapid exchange of larger amounts of data. Computing capability and data storage were other bottlenecks that initially limited the routine analysis to reported parameter data. In the beginning the Group envisioned more than one standardized international center. The Group saw the need for redundancy and the value of having more than one independent view in the analysis and was here influenced by the WMO, which had three centers. The Group also initially presented some estimates of expected detection and location capabilities based on simulated networks of 50 stations. These early estimates have proved to be in fair agreement with later experience, even if the seismic network in the CTBT is estimated to have improved capabilities, which we discuss later.

3.2.2 The First GSE Test

After its first report outlining the basic design, the Group continued its detailed work on specifications and methods, at times in a political atmosphere that was not conducive to rapid progress. The last of the following two reports by the Group in 1979 and 1984 (CD/43 1979, CD/448 1984) noted that the advantages of digitally recording seismological stations were becoming widely recognized and a growing number of such stations had been installed. As part of national efforts, experimental international data centers were developed in several countries and a number of small-scale data exchange tests were conducted. The first large-scale test was conducted in late 1984 and was named Group of Scientific Experts Technical Test 1 (GSETT-1). The purposes of that test were to test the extraction of parameter data at stations, the transmission of such data over the WMO GTS and the procedures for handling those data at Experimental International Data Centers (EIDCs). In all 75 seismological stations in 37 countries participated by contributing Level I data and three EIDCs were operating in Moscow, Stockholm and Washington D.C. The Group concluded that the test *"attracted a remarkably wide participation"*. The Group also concluded that a future global system should be equipped with digital recording systems. The experiment supplied a great deal of valuable experience related to the extraction and exchange of Level I data using WMO GTS. It was also realized that a lot of further work was needed to develop efficient procedures for data analysis.

3.2.3 The Second Test – Moving into the Digital Era

The second large scale test by the Group, GSETT-2, was conducted in 1991 (CD/903 1989, CD/1144 1992). 60 stations in 34 countries participated, together with four EIDCs in Canberra, Moscow, Stockholm and Washington D.C.

A considerable technical development had occurred since the first test seven years earlier. More than two-thirds of the stations could now record data digitally and an increased number of array stations were participating. National Data Centers (NDCs) were given an extensive role in the test to extract and provide both parameter and waveform data. This turned out to be a major undertaking for many NDCs. The communication links had also considerably improved their capabilities since the first test. Dedicated computer-to-computer links were established between the four EIDCs with capacities from 9.6 to

54 kbps, quite modest numbers by today's standards but quite impressive at the time. Moreover, a number of connections between NDCs and EIDCs were computer-to-computer connections using commercial links. The use of WMO GTS was gradually marginalized. The EIDCs were conducting their analysis independently of each other but could reconcile their interpretations. They also requested additional waveform data for a large fraction of the events to improve on event location and characterization.

3.2.4 GSETT-3: Approaching a Verification System

The third and last of the Group's large-scale tests, GSETT-3, marked yet another most significant step towards the final verification system (CD/1254 1994, CD/1423 1996). It was in fact a test of a prototype of the seismological component of the verification system to be adopted during the negotiations of the CTBT.

Based on the results of GSETT-2 and on the rapid development in communications and computer processing and memory, the concept had been developed further. There were two main developments: one was to have a two-tiered network of some 50 Alpha or primary stations, as they are called in the Treaty, and about 100 Beta or auxiliary stations. The other was to have just one single Experimental International Data Center. This center would receive only waveform data, now all in digital form, from the primary stations, which were continuously on-line, and from auxiliary stations on request. The EIDC of GSETT-3 was located in Washington D.C. GSETT-3 began full-scale operation on 1 January 1995 with the participation of 60 countries. The test involved 43 primary and 90 auxiliary stations. This experimental network contained 32 protocol to the CTBT and 38 of the 120 auxiliary stations. Based on the experience from the on-going GSETT-3 the Group was requested in 1995 to propose an auxiliary network to the on-going CTBT negotiations (CD/1372 1995). The auxiliary network proposed by the GSE is very close to the one finally adopted for the Treaty.

The test continued through the initial build-up of the facilities of the Provisional Technical Secretariat (PTS) of the CTBTO Preparatory Commission in Vienna and did not end until the Washington center closed its experimental IDC function in March 2000.

The initial analysis system in Vienna was essentially a copy of that developed in Washington and for some years a high capacity computer-to-computer link connected the centers in Washington and Vienna to facilitate the build-up of an analysis capacity at the PTS. Data transfer from the stations of the International Monitoring System (IMS) was gradually shifted from the Washington center to the PTS. This seamless transition from GSETT-3 to the IMS system greatly facilitated the build-up of the seismic component of the IMS.

3.2.5 GSE – Successful Preparatory Work

GSE was a thorough scientific and technical undertaking to develop and test a concept of a seismic verification system. The tests conducted by the GSE also demonstrated that the system could be implemented and operated and it also gave a good indication of the capabilities to be expected. For political reasons the GSE work was limited to the seismological component of the global verification system. Informal discussions on other technologies took place in connection with the GSE, however. In 1982 a radionuclide network of 100 stations was proposed, employing both particulate and noble gas sensors. This network resembles the inherent ambitions of the radionuclide network of the CTBT monitoring system. The GSE design of the seismic system proved also to be a model for the infrasound and hydroacoustic components of the CTBT monitoring system. The lack of preparatory scientific and technical work on these two components and on the system for detecting radioactive particles and noble gases became evident during the CTBT implementation phase, as will be discussed further later in this book.

Experts from a large number of countries participated in the work of the GSE, which thus became a mutual training activity and provided a common understanding of CTBT verification issues among international experts. In this way expertise was created that benefited national delegations taking part in the final negotiations. Many of the GSE experts have also played important roles in the implementation of the CTBT. Some have been active members of the technical Policy Making Organ, Working Group B, of the CTBTO Preparatory Commission responsible for the implementation of the CTBT. Others have joined the PTS, which supports the CTBTO Preparatory Commission in the technical implementation. GSE also helped build a culture of international technical cooperation. Through many joint projects, technical facilities and scientific capacities were also created in developing countries. Such national capabilities are essential for States to be able to utilize and benefit from the verification data provided by the PTS.

GSE members were appointed by governments and they were normally senior scientific experts in the field of seismology, belonging to national scientific institutions. This provided an important connection between government authorities and the scientific community, which proved to be essential to the success of the GSE as work could progress at two different and well connected levels: as a formal negotiation in Geneva and as informal cooperation among the many national scientific institutions that became involved. States provided substantial support to the work of GSE through contributions to their national institutions and to cooperative efforts, such as the build-up of facilities in developing countries. New monitoring stations were established and a large number of scientific studies were conducted through national initiatives encouraged by the GSE. Experimental International Data Centers were created specifically in support of the GSE work. Initially such centers were established in four countries and later efforts were concentrated on a large-scale prototype center in Washington D.C., having an international staff of some 50 persons.

GSE planned, conducted and evaluated three large-scale, global tests to study procedures for data collection and data analysis and to gain experience from the operation of a complex global verification system. These tests also demonstrated that such a system could be established and function in the real world, not just on paper. The tests also provided data that made it possible for each participant to estimate the verification capabilities that could be expected from such a system. This was essential information during the negotiations, when States assessed the proposed verification measures in the CTBT.

3.3 The CTBT Negotiations 1994–96

Work on the political level also continued in Geneva in the Conference on Disarmament (CD) (Fig. 3.2). A number of proposals were presented on a number of issues, in particular on different aspects of the difficult issue of on-site inspections. More comprehensive treaty proposals were also introduced. In 1983 Sweden proposed a draft treaty that provided for fact finding on-site

Fig. 3.2a Francisco de Vitoria Council Chamber in the Palais des Nations, Geneva. This is the room where CD assembles. The murals in this gloomy hall are by the Spanish artist José María Sert (1874–1945)

Fig. 3.2b Ceiling of the Francisco de Vitoria Council Chamber. This mural "Solidarity" represents the solidarity among States from all continents in achieving peace

inspections, and in 1987 a group of socialist countries, including the USSR, presented "Basic provisions of a treaty on the complete and general prohibition of nuclear weapons tests" (CD/756 1987). Among the provisions were mandatory on-site inspections that a State receiving a request must accept. In 1991 Sweden tabled a draft CTBT (CD/1089 1991) that further developed the on-site inspection provisions to include inspections by invitation to help clarify an event on one's own territory. A great deal of effort, commitment and hard work, but no real progress, is a good summary of the political work over some 30 years.

Again events outside the Palais des Nations in Geneva (Fig. 3.3) proved crucial. The 1990 NPT Review Conference failed to produce a consensus report, largely due to a lack of progress on a comprehensive test ban. At the 1991 PTBT Amendment Conference a large number of States called for CTBT negotiations to start without delay in the new political situation after the collapse of the USSR. More importantly, the main powers were ready to stop testing. The USSR declared a moratorium in 1990 and the USA in 1992 and, as a consequence, the UK did the same. In 1993 – finally, many will say – the CD established an Ad Hoc Committee on a Nuclear Test Ban and negotiations were about to start.

Ramaker et al. (2003) give a comprehensive review of the history of the CTBT negotiations. Our brief summary is based on that review and our own experience from the negotiations.

Fig. 3.3 View of the Palais des Nations. The Palais, built between 1929 and 1938, was intended to house all the organs of the League of Nations in Geneva. Behind the heavy gates and walls of this building, States can find the place "*to work, to preside and to hold discussions, independently and easily in the calm atmosphere which should prevail when dealing with problems of an international dimension*" (from assignment to the architects)

3.3.1 Starting the Negotiations

On 10 August 1993 CD (CD/1212 1993) gave its Ad Hoc Committee on a Nuclear Test Ban a mandate to negotiate a CTBT. It further requested the chairman of the Ad Hoc Committee, Ambassador Y. Tanaka of Japan, to consult "*on the specific mandate for, and the organization of, the negotiation*". In December 1993 Tanaka reached a consensus on such a mandate, which was formally adopted in January 1994 (CD/1238 1994). To agree on the method of work proved more difficult, as did the issue of who should chair the Committee during 1994. Eventually the Mexican Ambassador Miguel Marin-Bosch was appointed and negotiations commenced on February 3, 1994. The way to organize the negotiations was quickly agreed. As early as February 16, two Working Groups were established, one on verification chaired by the German Ambassador W. Hoffmann and one on legal and institutional matters chaired by Ambassador L. Dembinski from Poland. During 1994 four Friends of the Chair were appointed to deal with specific issues in the verification area: the seismic technique, non-seismic techniques, on-site activities and transparency measures. For the other Working Group, two Friends of the Chair were appointed, for the Organization and for Entry Into Force. A connection was also established between the Working Group on Verification and the Group of Scientific Experts. At the start of the negotiations CD had 38 member States, growing to 61 in 1996, while several additional non-members could participate in the negotiations, but were not allowed to block consensus in the CD. Israel, New Zealand and South Africa were non-members that participated actively.

As early as January 5, 1994, Australia submitted a Treaty outline (CD/1235 1994) to be used as a basis for the negotiations. In the summer of 1994 there was a spirited debate about whether or not the chairman should already present a chairman's "vision" text. Eventually it was concluded that it was too early and instead the CD Secretariat composed a "rolling text" containing the status of the considerations so far in the two Working Groups. This text was annexed to the report submitted at the end of the 1994 CD session in September to the UN General Assembly (A749/27 1994). The Ad Hoc Committee held a number of intersessional meetings in November and December, 1994, focusing on verification issues and the future Organization to be established.

3.3.2 1995 – Waiting in the Hallway

CD has a method of work mandating that everything has to be re-created each year and the chairpersons have to rotate among the States. The Ad Hoc Committee was re-established at the outset of the 1995 CD session and Ambassador Dembinski from Poland was appointed chairman. It took more efforts and time to decide that the Swedish Ambassador Lars Norberg should chair the Working Group on verification and Ambassador Jaap Ramaker of the Netherlands should chair the Working Group on legal and institutional issues. Ten new Friends of the Chair and two so-called Convenors were appointed on particular issues in a similar way as in 1994. One of the Friends of the Chair was assigned the task of preparing for the Preparatory Commission of the future CTBT organization and a draft text on its establishment was prepared during 1995.

The technical verification issues came into the foreground and a number of experts worked successfully on a broad spectrum of issues. Peter Marshall played a key role in bringing a lot of these issues together with his "tiger teams", to identify the different global networks that would form the global monitoring system in the CTBT, referred to as the International Monitoring System (IMS). Achieving a common understanding on these issues was a significant step forward. On the political side a common understanding was reached on the scope of the Treaty when all States supported a "zero yield comprehensive test ban" with no reservations.

The NPT Review and Extension Conference held in April – May 1995 pronounced that "*the completion by the Conference on Disarmament of the negotiations on a universal and internationally and effectively verifiable Comprehensive Nuclear-Test-Ban Treaty no later than 1996*" is an important measure for "*the full realization and effective implementation of article VI*" of the NPT (UNODA 1995).

After the conclusion of its 1995 session, CD again reported to the UN General Assembly (CD/1364 1995), this time including an even more extensive rolling text, now containing the list of the primary seismic stations that the GSE had been working on for so many years. At their summit on October 23, 1995, Presidents Clinton and Yeltsin announced that the Russian Federation and the USA would "*work together to succeed in getting a zero-yield comprehensive test ban treaty*" in 1996 (Office of the Press Secretary of the White House 1995). To increase pressure on the negotiations the USA announced at the beginning of the 1995 negotiations that the USA would extend its moratorium on testing until the entry into force of a CTBT, provided the Treaty would be signed before September 30, 1996. Furthermore, a UN general Assembly resolution (A/RES/50/64 1995) urged States to conclude a CTBT "*as soon as possible and no later than 1996*".

3.3.3 1996 – The Endgame

The final negotiations commenced on 21 January, now with the Dutch Ambassador Jaap Ramaker as chairman and as before with two Working Groups, the Russian Ambassador G. Berdennikov chairing the one on verification and Ambassador M. Zahran of Egypt chairing the other. The number of supporting Friends of the Chair and Moderators had now increased to 17 in all. The negotiations resumed on the basis of a rolling text filled with an increasing number of brackets. In addition to that rolling text, in February the Islamic republic of Iran presented a draft treaty (CD/1384 1996) and Australia a model treaty text (CD/1386 1996).

When Ramaker announced in early March that he would soon present a paper with an outline of a draft treaty, the negotiations changed gear and interest focused on the chairman's expected text. The outline was presented on March 20 and after a number of consultations Ramaker announced that there would not be a further version of the rolling text but rather a draft Treaty. The draft (CD/NTB/WP.330 1996) was presented on May 28. It contained 88 pages without brackets, where he had tried to capture consensus and bridge differences and above all to "*harmonize the desirable with the attainable*". There was a mixed reaction to the chairman's text and some delegations would not accept it as a basis for further negotiations, preferring to stay with the rolling text. A number of issues were creating problems for delegations; such as the preamble, where India wanted reference to nuclear disarmament in a time-bound framework; and the Organization, in particular the Executive Council, where the number of seats was increased from 45 to 51 in the final rounds of negotiations. The introduction of six rather than the normal five regional groups in the Executive Council also created difficulties for some States. The entry into force clause was one of the more difficult ones, listing 44 States with nuclear activities that have to ratify before the Treaty can enter into force. India resisted this idea and suggested a model used earlier, e.g. in the Chemical Weapons Convention, requiring 65 ratifications for EIF. India also refused to have any IMS station on its territory named in the Treaty and the "*to be determined*" that can be found in the list of stations in the Treaty refers to stations that were placed in India in earlier versions of the draft Treaty text. After an intensive period of meetings and consultations, Ramaker presented a revised draft on 28 June, the closing day of the second part of the CD session. When the Ad Hoc Committee resumed its work on 29 July a number of States, including the Russian Federation and the USA, stated that they were prepared to accept the draft. Others, in particular India and Iran, voiced their opposition. After a final revision of the rule on how to take decisions in the Executive Council and some relatively minor corrections a chairman's revised text (CD/NTB/WP.330/Rev.2 1996) was presented on 14 August.

This was the final text, the question now was how to proceed. The Ad Hoc Committee on a Nuclear Test Ban took its decisions with consensus and it was clear that there was no agreement to transmit the text as it was from the Ad Hoc Committee to the CD itself. All the Ad Hoc Committee could do was present a report containing a description of the negotiations and the chairman's conclusions on his consultations. The Treaty text that enjoyed an overwhelming support by States participating in the negotiations was thus held hostage in the Ad Hoc Committee. What now? The only way forward was through national initiatives. The chairman's text of August 14 was presented on August 22 to the CD as a national paper by the Belgian delegation (CD/1427 1996). The same text was then introduced by Australia as a draft resolution to the UN General Assembly, where it was adopted on September 10, 1996. Decisions in the General Assembly can be taken by vote and only three States – Bhutan, India and Libya – voted against, five abstained and 158 voted in favor. On 24 September 1996 the CTBT was opened for signature at the UN Headquarters in New York. We had reached the end of a long journey. Few expected that we were at the start of another long journey, which still has to reach its destination.

3.4 Critical Issues During the Negotiations

During the negotiations a number of issues turned out to be more critical or controversial than others. A number of those issues have also proved to be in focus during the CTBTO Preparatory Commission's work. A detailed review of the negotiations of the individual elements of the Treaty is given by Ramaker et al. (2003). The intent here is only to briefly mention a few of these issues, on which much effort and argument was spent.

3.4.1 The Preamble

The preamble to a treaty is where States try to make references to broad goals and additional desired activities. The main discussion was on a proposal to see the CTBT "*within the framework of an effective nuclear disarmament process*". This proposal was rejected and the preamble ended up with a more general formulation "…*constitute a meaningful step in the realization of a systematic process to achieve nuclear disarmament*".

3.4.2 The Organization

In the negotiation on the CTBT Organization two issues were at the forefront: would a new independent organization be established or could the IAEA be entrusted with the task? Secondly, how would the Executive Council be composed and how would it take decisions? It became clear fairly early that Vienna would host the Organization, and also that it would be an independent organization with no special ties to the IAEA. As we have already, seen the Executive Council was a harder nut to crack and an agreement was reached only at the last minute. There were basically two questions: How many members should the council have? And how many regional groups should there be: five or six? Agreement was finally reached on 51 members and six regional groups, one more than in the Chemical and Biological Weapons Conventions. This additional group was to cover the Middle East and South Asia. As we shall see later, this brought the Middle East conflict directly into the Treaty and to the work of the CTBTO Preparatory Commission.

3.4.3 Verification

CTBT has the most elaborate international verification system ever created and a large portion of the Treaty text and an extensive protocol deal with verification issues. Verification-related issues also took up a large part of the negotiations, with the GSE preparatory work starting as early as 1976. The verification discussion included three main issues: which technologies should be used and what the networks of monitoring stations would look like; how far the Organization should carry the analysis of the data; and finally, issues related to on-site inspections.

To start with the technologies, it was clear from the outset that a global seismic system would be a key component of the international monitoring system. GSE had worked on such a system for almost 20 years and conducted global tests. The negotiators could quickly agree on the design of the seismic system, even if a lot of detailed issues took some time to settle. The main issue here was which other technologies should complement seismic monitoring. In addition to measurements of radionuclide particulates and noble gases, systems for observing hydroacoustic and infrasound signals also made their way into the Treaty. The general understanding of what could be achieved by each of these three technologies was quite limited. The cost estimates that were floated during the negotiations were also gross underestimates.

Satellite observations as well as optical and electromagnetic pulse (EMP) observations were also considered. A few countries had some knowledge of all those technologies, but very little knowledge was shared internationally and no broad international studies or tests had been carried out, similar to those in the seismic field. At this time satellite observations were not generally available, they were rather expensive and suffered from limited coverage and resolution, which may explain why such observations were not included. A specific satellite system established for test ban verification was considered, with a frightening price tag. Many felt that satellite observations should rather be part of national technical means. The knowledge of and experience with ground-based optical and EMP observations were too limited to attract enough interest.

A most important issue was: what is the role of the Organization when it comes to verifying compliance with the Treaty? It became clear fairly early that States did not want the Organization to make any final assessments as to the nature of the events that were observed. Such decisions should rest with the States. It was, on the other hand, a general understanding that the Organization should facilitate national interpretation by providing a user-friendly bulletin of all events. It was agreed that the Organization should detect and locate any event that could be discerned from the observations and provide all States with that information and all raw data collected. A key question was how far the Organization should go in pointing out events that might depart from what could be considered natural events. This issue of screening, which is also difficult from a technical and methodological point of view, was never fully resolved. The text in the Treaty is also quite vague on which screening procedures to apply and this issue is still being considered in the CTBT implementation phase.

The discussions of on-site inspections (OSIs) were conducted on two levels, political and technical. The political level was very much concerned with the procedures for taking a decision to initiate an OSI. This discussion was carried out using the concepts of "Red Light", where an OSI would take place unless the Executive Council stopped it, and "Green Light", where an OSI would not commence until the Executive Council approved the request. Another closely connected issue that took long time to resolve was the number of affirmative votes needed in the Executive Council. Finally the outcome was "Green Light" and 30 affirmative votes. Another political issue was the maximum size of the inspection area: should it be 1000 km^2, as was finally agreed, or only 100? The role of a consultation and clarification process in relation to initiating an OSI was another issue; should it be required or only encouraged, as finally incorporated in the Treaty. Yet another political issue related to the use of information from national technical means as a basis for an OSI, and it was eventually agreed that such information can be used as a basis for an OSI request.

On the technical level, procedures, methods and equipment to be used in the conduct of an OSI were elaborated in great detail. These negotiations were complicated by the fact that few States had done any studies or experiments related to this issue and the generally shared knowledge was quite limited. This applied both to the methodology of conducting an OSI over a large area and the equipment to be used. The negotiators agreed on an overall structure on for the conduct of an OSI, on time lines for an inspection, and on the technologies to be used. A lot of specific technical details were also included in the protocol of the Treaty without too much background, and these details are further developed in the CTBTO Preparatory Commission's work on the implementation of the OSI regime.

3.4.4 Entry into Force (EIF)

The most controversial article in the CTBT is the one on the formula for entry into force. The ambition to include all States of concern gave each of them the possibility to take the Treaty hostage. How did it happen that the article was created this way? From the beginning many delegations expected to have a simple provision stipulating that a certain number of States are required to ratify for EIF, in a way similar to the Chemical Weapons Convention (CWC), which requires 65 ratifications. The experience from the CWC, which at this time was approaching 65 ratifications without including the two major possessors of chemical weapons, gave several States cause for concern. Most delegations agreed that it was important to get all nuclear weapons States and some nuclear "threshold" States on board from the very beginning, but many were initially hesitant about a complicated EIF formula. As time progressed a number of proposals were developed to find a good way of describing the desired goal without identifying a limited number of States. Some suggested all CD members, others all CD members plus all those States having nuclear reactors known to IAEA. Another proposal suggested, maybe surprisingly, that EIF should require ratification by all States having primary seismic stations and radionuclide laboratories. From time to time warnings were expressed that any list of States would create the potential for hostage taking. In the end, the list now appearing in Annex 2 to the Treaty was proposed in the chairman's text, containing those CD members that possessed nuclear reactors at the time the negotiations were concluded.

3.5 Reflections on the Negotiations

The negotiation of the CTBT turned out to be a major undertaking over a fairly long period of time and with many participating delegations. Many delegations and individuals put in a lot of dedication, effort and creativity to move the process forward. The four chairmen of the Ad Hoc Committee – Ambassadors Tanaka, Marin-Bosch, Dembinski and Ramaker – can take due credit for their stewardship. Ambassador Ramaker and the Dutch delegation displayed great leadership and put in a large amount of work and ingenuity to successfully conclude the negotiations. Verification issues were an important part of the negotiations and Peter Marshall of the UK, with his exceptional skills in communicating technical matter to diplomats, played a key role in bringing the technical matter into the Treaty. Jenifer Mackby was the secretary of the negotiations and her meticulous efforts greatly served the negotiations.

The negotiations were successful; agreement on a CTBT was reached. The proof of the pudding is, as always, in the eating and so far the dessert is still in the pantry, waiting to be served. It is too early to tell if the Treaty will turn out to be a good one. Later in the book the reader will find a few comments on different parts of the Treaty as we have experienced it so far. Looking at the negotiation process, it may be said that it took some forty years for the CTBT issue to mature, but once the real negotiations started agreement was reached in two and a half years. Compared to the couple of weeks it took to negotiate the PTBT or the few months it took to conclude the TTBT, this might be considered a long time for concluding an agreement on such a specific issue as banning nuclear testing. Compared to other multilateral negotiations, on the other hand, the time is quite short. Multilateral negotiations generally take a long time as a large number of delegations are involved, having their own national positions that have to be taken into account. Any issue related to nuclear weapons is also sensitive, especially to the nuclear weapon States. The CTBT negotiators also had to agree on unprecedented verification provisions, including several technologies about which there was little shared knowledge or experience among the parties. In the final part of this chapter we reflect on experiences from the GSE and on the ways the CD as an institution affected the negotiations.

3.5.1 GSE a Tool to Get Prepared

CD took an unprecedented decision when as early as 1976 it established GSE and gave it a long-term mandate to develop and test a seismic verification system. GSE proved that it was possible and most useful to conduct preparatory technical work prior to political negotiations. States were ready to contribute experts, build high quality monitoring stations and engage not only in numerous scientific studies but also in testing on a global scale. The work of the GSE was not seen by anyone as a substitute for political negotiations nor a commitment to future such negotiations. It was a thorough scientific and technical effort to develop and test a concept of a seismic verification system. The tests conducted by the GSE also demonstrated that the system could be implemented and operated and it also gave a good indication of the capabilities to be expected.

A number of very basic circumstances made the GSE work successful: one was that GSE could provide for a sustained and focused efforts over a long period of time. This was possible as the CD at an early date gave GSE a long-term mandate that did not have to be renewed every year, as was the case with other subsidiary bodies of the CD. Furthermore, the GSE was not obliged to change its chair on a monthly or yearly basis. In fact GSE had only two chairmen during the entire twenty-year period. GSE also had a Scientific Secretary who served for the entire period. These conditions were all conducive to focused, well-planned activities. Based on its mandate GSE established a sustained multi-year agenda that provided for activities in the meeting rooms in Geneva but also for cooperation among scientists and institutions around the world. Almost all experts in the GSE were seismological scientists. It might prove useful to include experts on systems analysis and engineering in similar work in the future.

The experiences from the GSE work provide good arguments to explore, at an early stage, technical and other non-political issues related to possible future treaties. Expert elaborations are in no way substitutes for political negotiations but ways to prepare the ground for political negotiations that might follow and for the eventual implementation of a treaty. Such work could be conducted within frameworks that do not in any way imply that political negotiations might follow or prejudge their possible outcome. It could also help create mutual confidence and enhance international cooperation in the actual fields. It would help increase and share knowledge among experts around the world of what can be achieved technically. It might also encourage States to establish infrastructure that might prove important in the implementation of possible future treaties.

3.5.2 CD a Bottleneck?

The CTBT negotiations during 1994–1996 illustrate two sides of the CD. On the up side it demonstrated that CD could manage a technically complex negotiation and provide a competent and efficient secretariat. CD also managed to mobilize the energy and engagement of many delegations in a negotiation lasting over several years. The downside was clearly illustrated by the need to bring the text from the negotiating Ad Hoc Committee to the CD Plenary and subsequently to the UNGA through national initiatives. CD failed, due to its consensus procedure, to properly conclude the negotiations and provide a treaty ready for signature.

Since the CTBT was negotiated in 1996 no treaty negotiations have commenced in the CD for over ten years. Why did the CTBT get stuck in the end and why has nothing happened thereafter in a period when international relations among nations have been most relaxed? Is the CD facilitating multilateral disarmament negotiations or is it a bottleneck? It is the only forum dedicated by the UNGA to multilateral disarmament negotiations. It has a permanent secretariat and well suited localities in the prestigious Palais des Nations to support negotiations. 65 States have permanent delegations present, which are meeting regularly but making no progress. How come? Many will argue that there is a lack of political will to make progress. This is most likely true, but is it the only reason for not even getting started?

CD has a number of built-in features that may prevent it from being an efficient negotiating body. It is a large assembly. The membership in what today is the Conference on Disarmament has grown gradually from the original 10 members of the UN Ten Nation Committee on Disarmament in 1960 to the present 65 members, or about a third of all UN member States. This increase has broadened the political and geographical representation and it might also be positively interpreted as a growing international interest in disarmament issues. The increased membership has made it increasingly more time consuming to make progress. CD is no longer a small, efficient negotiating body able to provide disarmament initiatives. In a way it has become a "mini UN", but without the ability to take majority decisions. Maybe an even greater hurdle to progress is that the CD has maintained the concept of regional groups from the Cold War era. The main powers still have a pronounced influence on the progress or lack thereof.

CD operates by consensus. This means that any of the 65 States can keep any issue hostage for any length of time. This was clearly illustrated in the final stage of the CTBT negotiations, as discussed above. The consensus rule applies not only to issues of substance but also to procedural issues such as the meeting agenda and the creation of groups to deal with specific issues. CD has to re-establish its agenda and, in particular, its work program every year, which provides States ample opportunities to obstruct proceedings and prevent any actions on substance. This has also been the case for more than ten years. States have also established linkages between very different issues and refused to deal with one without simultaneous action on another.

The chair of the CD rotates among delegations in alphabetic order every month when the CD is in session. This means that all States share responsibility for the activities. With a membership of 65 States and six chairmanships per year, a delegation will be in the chair for one month less than once in ten years. You may debate the kind of engagement this creates. The rotating chairmanship prevents strong leadership and makes it difficult to focus on long-term perspectives. It is also difficult to conduct the necessary consultation processes to create conditions for new negotiations if you have to pass the torch from one ambassador to the next every month.

CD has established a routine bureaucratic procedure for doing its day-to-day business. It holds regular meetings twice every week where delegations may make statements on anything they see fit. From time to time high level political persons visit to show support. This routine procedure, which also includes reaching agreement on an annual report to the UNGA, may give the impression of ongoing activities and masks the fact that nothing substantial is being achieved.

The "Mine Ban Treaty" was negotiated in 1997 outside CD in an ad hoc way among those interested, with the close engagement of non-governmental organizations (ICBL 2007) and smaller States. This treaty was negotiated within one year and the number of signatories has gradually increased, reaching 155 States by the autumn of 2007. This is an interesting alternative way of conducting disarmament and arms control negotiations, focusing on a particular issue and leaving aside the heavy backpack of bureaucratic procedures.

At the 2005 UN General Assembly First Committee a group of six States floated a draft resolution that called for the creation of four ad hoc committees under the UNGA to engage in substantive discussions on issues including a fissile material ban (Estabrooks 2006). Lacking the needed support, the proposal was eventually dropped. If CD continues to show no progress, States might continue to search for new, potentially more efficient forums for negotiating or preparing for possible future treaties. To meet this challenge the CD must develop more efficient procedures for conducting its work or face the risk of becoming obsolete.

Chapter 4
The Treaty

International treaties and agreements are playing an increasing role in promoting international security in an environment where non-military elements are becoming increasingly important. Today efforts to preserve our global environment and to prevent the proliferation of weapons of mass destruction are at the head of the international security agenda. The CTBT is an essential contribution to the global non-proliferation regime and a step towards nuclear disarmament. More than ten years have now passed since the Treaty was opened for signature and it has not yet entered into force for reasons we discuss later. All States Signatories, including those that have not ratified, have declared their intent to renounce further testing. Even though it is not in force, the CTBT has thus established a norm of non-testing among its signatory States, which is politically difficult and costly to break.

India and Pakistan, which have not signed the Treaty, have also declared unilateral moratoria on testing after their test series in 1998. The Indian Prime Minister stated before the UN General Assembly that his country will not stand in the way of entry into force of the Treaty (Subrahamanyam 1999). It is generally assumed that the same holds for Pakistan. A significant step towards a resolution of the nuclear situation in North Korea was taken by the agreements reached on February 13, 2007 (Feffer 2007) during the six party negotiations involving North Korea, South Korea, China, the Russian Federation, Japan and the USA.

In this chapter we present and discuss important parts of the CTBT. The full text of the Treaty is available on the CTBTO Preparatory Commission's home page (CTBT 1996).

4.1 The Preamble

The preamble of a treaty normally expresses some general considerations and commitments related to the actual treaty. So too the CTBT. The preamble clearly links the Treaty to efforts at achieving nuclear non-proliferation and disarmament. The States Parties are "*convinced that the present international*

O. Dahlman et al., *Nuclear Test Ban*, DOI 10.1007/978-1-4020-6885-0_4,
© Springer Science+Business Media B.V. 2009

situation provides an opportunity to take further effective measures towards nuclear disarmament and against the proliferation of nuclear weapons in all its aspects and declaring their intention to take such measures". The States Parties are *"stressing therefore the need for continued systematic and progressive efforts to reduce nuclear weapons globally, with the ultimate goal of eliminating those weapons"*. States are thus in general terms declaring their intention to take effective measures towards nuclear disarmament and nuclear non-proliferation by reducing nuclear weapons globally. This was in 1996 and not much, if anything, has been delivered on those promises.

The preamble also recognizes the CTBT as a meaningful step en route to nuclear disarmament. States Parties *"recognizing that cessation of all nuclear weapon test explosions and all other nuclear explosions, by constraining the development and qualitative improvement of nuclear weapons and ending the development of advanced new types of nuclear weapons, constitutes an effective measure of nuclear disarmament and non-proliferation in all its aspects"*. States are *"further recognizing that an end to all such nuclear explosions thus constitute a meaningful step in the realization of a systematic process to achieve nuclear disarmament"*. States thus acknowledge that the CTBT will constrain the development of nuclear weapons and that this is one of the purposes of the Treaty and not an undesired consequence.

So is the CTBT primarily a disarmament or a non-proliferation treaty (Schaper 2007)? In our view it is a bit of both. And despite all technicalities that will be discussed below, its main impact is political. It is an important building brick in the non-proliferation regime and an essential element in promoting nuclear disarmament. It is a treaty that was discussed and negotiated from time to time over 50 years before it was finally agreed. The very fact that it proved possible to agree on a CTBT might be the most important consequence and opens the way for other treaties to follow suit. This is one of the reasons why it is so important that the Treaty formally enters into force, even if nobody anticipates that any State Signatory will resume testing.

4.2 Basic Obligations

The Treaty has a straightforward purpose and *"each State Party undertakes not to carry out any nuclear weapon test explosion or any other nuclear explosion, and to prohibit and prevent any such nuclear explosion at any place under its jurisdiction or control"*. States also undertake to *"refrain from causing, encouraging, or in any way participating in the carrying out"* of any nuclear explosion.

The Treaty contains no definition of a nuclear explosion. In general there is an understanding of what constitutes a full-scale nuclear explosion. The controversy has developed around two kinds of tests, subcritical and hydrodynamic, aimed at studying the behavior of nuclear material under very high temperatures and pressures. Sub-critical tests (Global Security 2007) are experiments where small amounts of fissile material are used but no chain reaction occurs. In hydrodynamic tests the fissile materials are replaced by non-fissile ones. Such tests have been an element in the development and maintenance of nuclear weapons over a long period of time. A number of such tests have also been carried out since the CTBT was signed. This has created concern among some States and Non-Governmental Organizations, which also want those tests to be stopped. There might be arguments about why such sub-critical and hydrodynamic tests should be stopped as part of further efforts to limit the development and the future reliance on nuclear weapons. We, however, think this should be a separate discussion that should not be undertaken in relation to the CTBT, as this Treaty and its verification arrangements were not aimed at such experiments but rather at explosions generating a substantial amount of nuclear energy.

In the past, nuclear explosions have, as discussed in our first chapter, been used for civilian and peaceful purposes, with few useful results. Article VIII, on review of the Treaty, deals with this issue. It states that a Review Conference, which might be held ten years after EIF, could, if requested, "*consider the possibility of permitting the conduct of underground nuclear explosions for peaceful purposes*". A decision to permit such nuclear explosions shall be taken by consensus. It is most unlikely that the issue of peaceful nuclear explosions ever will come up again unless some totally unforeseen need arises.

4.3 The Organization

An elaborate organization, the Comprehensive Nuclear-Test-Ban Treaty Organization (CTBTO), is to be established in Vienna to manage the Treaty. It consists of the Conference of the States Parties, the Executive Council and the Technical Secretariat.

4.3.1 The Conference

The Conference is the principal decision making organ for the Organization and will meet annually unless otherwise decided. The initial session of the Conference shall be convened by the Depositary, the Secretary General of the UN, no later than 30 days after the entry into force of the Treaty. All States Parties to the Treaty are members of the Organization and entitled to participate in the Conference.

The Treaty states that "*the Conference shall be the principal organ of the Organization. It shall consider any questions, matters or issues within the scope of this Treaty*". It further states that "*the Conference shall oversee the implementation of, and review compliance with, this Treaty and act in order to promote its object and purpose*". The Conference shall decide on the annual work program and budget of the Organization and on the scale of financial contributions to be paid by States Parties. It shall further elect members of the Executive Council and appoint the Director-General (DG) of the Technical Secretariat. The Conference shall also "*consider and review scientific and technological developments that could affect the operation of this Treaty. In this context, the Conference may direct the DG to establish a Scientific Advisory Board to enable him or her, in the performance of his or her functions, to render specialized advice in areas of science and technology relevant to this Treaty to the Conference, to the Executive Council, or to States Parties*".

4.3.2 The Executive Council

The Executive Council is the executive organ of the Organization and is responsible to the Conference. The Executive Council shall "*promote effective implementation of, and compliance with, this Treaty*". It shall further "*supervise the activities of the Technical Secretariat*" and consider the draft annual program and budget for the Organization. It shall also "*cooperate with the National Authority of each State Party*". The politically most important tasks of the Executive Council are to approve or reject requests for on-site inspections and to "*consider any concern raised by a State Party about possible non-compliance with this Treaty and abuse of the rights established by this Treaty*".

The Executive Council consists of 51 members elected for a term of two years from the six geographical regions defined in the Treaty. The Treaty defines the number of representatives from each region as follows: Africa 10, Eastern Europe 7, Latin America and the Caribbean 9, the Middle East and South Asia 7, North America and Western Europe 10, and South-East Asia, the Pacific and the Far East 8. The definition of these regions differs from that in the Convention on Chemical Weapons (CWC), which has five regions. The Asian region in the CWC is divided into two in the CTBT: one for the Middle East and South Asia and one for South-East Asia and the Pacific. The composition of the Executive Council was an important issue and detailed, you may say complicated, rules for election were established. "*At least one-third of the seats allocated to each geographical region shall be filled, taking into account political and security interests, by States Parties in that region designated on the basis of the nuclear capabilities relevant to the Treaty as determined by international data as well as all or any of the following indicative criteria in the order of priority determined by each region:*

(i) *Number of monitoring facilities of the International Monitoring System;*
(ii) *Expertise and experience in monitoring technology; and*
(iii) *Contribution to the annual budget of the Organization*".

4.3.3 The Technical Secretariat

The Technical Secretariat shall support the Conference and the Executive Council in their activities. It shall further prepare the draft program and budget for the Organization and the draft report of the Organization on the implementation of the Treaty. In the Treaty the Secretariat is given a number of defined responsibilities with regard to verification of compliance with the Treaty. This includes the day-to-day operation of the International Monitoring System and the International Data Center. The Secretariat shall also prepare for and support the conduct of an on-site inspection should there be a decision by the Executive Council to conduct such an inspection.

The Treaty states that "*the Technical Secretariat shall comprise a Director-General, who shall be its head and chief administrative officer, and such scientific, technical and other personnel as may be required. The Director-General shall be appointed by the Conference upon the recommendation of the Executive Council for a term of four years, renewable for one further term*". The staff shall be chosen "*securing the highest standards of professional expertise, experience, efficiency, competence and integrity*". Only citizens of States Parties can serve in the Organization and "*due regard shall be paid to the importance of recruiting the staff on as wide a geographical basis as possible*".

4.4 Verification

A unique, most interesting part of the Treaty is its extensive verification regime, containing four elements: an International Monitoring System, provisions for consultation and clarification, on-site inspections, and confidence-building measures. Each State Party undertakes to cooperate with the Organization and other States Parties through a designated National Authority to facilitate the verification of compliance with the Treaty. In addition to using this verification regime "*no State Party shall be precluded from using information obtained by national technical means of verification*".

4.4.1 The International Monitoring System

The purpose of the International Monitoring System (IMS) is to provide States with equal, high quality information for national assessments. This is a way, at least in principle, to give all States equal opportunities to verify compliance with the Treaty. The CTBTO is a service organization to the States. It provides each State with data obtained through analysis of all available observations on a routine and timely manner and any raw data a State may request. The CTBTO does not have the authority to draw any conclusions as to the nature of the events it observes. In this respect its tasks differ considerably from those of IAEA in relation to the NPT. It is up to the States to conclude if an observed event might need further clarification. If that is the case the State may request clarification and in the end an on-site inspection. The general principle of making it possible for each State to participate in the monitoring of a treaty on an equal footing by establishing comprehensive international verification measures could also prove useful for possible future treaties.

The International Monitoring System is composed of a total of 321 monitoring stations in four technologies: seismological, hydroacoustic, infrasound and radionuclide; 16 radionuclide laboratories and an International Data Center. The seismological system is a two-tiered system with a primary global network of 50 stations, reporting all data on-line to the International Data Center (IDC), and an auxiliary network of 120 stations from which data are requested by the IDC or States when needed. The hydroacoustic network, covering the oceans, has only 11 monitoring stations. Six of those are complex underwater installations and five are seismic stations placed on islands to record seismic signals converted from hydroacoustic signals in the ocean, so-called T-phase stations. The infrasound network comprises 60 stations distributed globally. Furthermore, the data from the infrasound and the hydroacoustic stations are reported to the IDC on-line. The radionuclide monitoring network is also two-tiered. It comprises a global network of 80 stations monitoring for radionuclide particles and 40 of those stations should also be able to monitor for radioactive noble gases upon entry into force of the Treaty. The radionuclide observations are transmitted to the IDC every 2 h. 16 radionuclide laboratories are also identified in the Treaty to support the monitoring stations in the data analysis. A description of the different verification technologies can be found in Chapter 2.

The IMS is under the authority of the Technical Secretariat, which has the responsibility to supervise, coordinate and ensure the operation of the system. The operation and maintenance of all IMS stations, except those in the auxiliary seismic network, are financed by the Organization. Operation and maintenance of the auxiliary seismic stations are paid for by the host States. The individual stations shall be owned and operated by the States hosting them. The Treaty contains provisions for a State to pay for the establishment of a station and then be compensated by a corresponding reduction in its assessed financial contribution.

States may, at their own expense, establish additional stations, referred to as cooperating national facilities, to contribute supplementary data to the International Data Center. Such stations must meet the same operational requirements as the IMS stations.

The International Monitoring System was not designed to meet any preset functional requirements in terms of capabilities to detect and locate events. The system is rather defined by its physical components, the 321 stations, with the provisional locations of each station given in Annex 1 to the Protocol to the Treaty. During site surveys to find a logistically suitable location for each station the coordinates were modified, some just in a minor way and others quite substantially. Annex 1 to this book gives the actual coordinates for stations as they are presently available. These modifications, which have to be approved by the first Conference of the States Parties, will, however, not affect the overall capabilities of the monitoring networks.

The International Data Center (IDC) "*shall receive, collect, process, analyse, report on and archive data from International Monitoring System facilities*". The IDC shall provide States with different compilations of the results of its analysis: a bulletin based on the automatic analysis and a more comprehensive bulletin based on human analysis of the observations from the IMS station network. These bulletins contain information on the location and the strength of the events detected. The protocol also requires that the IDC provide a "screened event bulletin" where standard event screening criteria have been applied to highlight "*events considered to be consistent with natural phenomena or non-nuclear, man-made phenomena*". An annex to the protocol provides some vaguely defined screening criteria. The IDC shall, if requested by a State, also apply national event screening criteria to create a special product for that State. The IDC should also provide each State with any amount of raw data it may request.

The Treaty contains procedures for modifying the International Monitoring System using a simplified amendment procedure. The Treaty tasks States Parties to examine "*the verification potential of additional monitoring technologies such as electromagnetic pulse monitoring or satellite monitoring, with a view to developing, when appropriate, specific measures to enhance the efficient and cost-effective verification of this Treaty*". The Treaty has some vaguely defined procedures to "*review scientific and technological developments that could affect the operation of this Treaty*". Ten years after entry into force a conference shall be held to review "*the operation and the effectiveness of this Treaty*". "*Such review shall take into account any new scientific and technological developments relevant to the Treaty*".

States further undertake to facilitate exchange related to verification technologies "*to enable all States Parties to strengthen their national implementation of verification measures and to benefit from the application of such technologies for peaceful purposes*". It is also noted that "*the provisions of this Treaty shall not be interpreted as restricting the international exchange of data for scientific purposes*".

4.4.2 Consultation and Clarification

The Treaty promotes the procedure of consultation and clarification to clarify issues related to possible non-compliance and it states: "*Without prejudice to the right of any State Party to request an on-site inspection, States Parties should, whenever possible, first make every effort to clarify and resolve, among themselves or with or through the Organization, any matter which may cause concern about possible non-compliance with the basic obligations of this Treaty*". The Treaty also describes the procedures that should be used if a State Party wants to engage the Director-General and the Executive Council in the clarification process.

4.4.3 On-Site Inspections

The Treaty also provides for on-site inspections to clarify observed events that cannot be confidently identified using other technical means or through consultations. Each State Party has the right to request an on-site inspection based on information collected by the IMS or by national technical means of verification, or a combination of these. The request shall be presented to the Executive Council and the Director-General. The Executive Council has to take a decision within 96 h after the receipt of a request and a decision to proceed with an on-site inspection requires at least 30 affirmative votes among the 51 members of the Executive Council. An on-site inspection shall have a duration of no more than 60 days and the Executive Council shall be provided a progress report within 25 days of the start of an inspection. The Executive Council may then decide to terminate the inspection. The Executive Council may, on the other hand, at the request of the inspection team, extend the inspection's duration by a maximum of 70 days in addition to the initial 60 days. An on-site inspection is to be carried out by inspectors from States Parties supported by the Technical Secretariat. The factual results of an on-site inspection are reported to States Parties and to the Executive Council. The Executive Council shall review the report of the inspection to address "*whether any non-compliance with the Treaty has occurred*".

Inspections could cover an area up to 1000 km^2, with no linear distance greater than 50 km in any direction. The inspected State has the right to declare restricted-access sites, each with an area no larger than 4 km^2. In all, 50 km^2 can be declared restricted-access sites. Such sites should be avoided by the inspection team but access can be requested and shall be granted to "*accomplish specific tasks within the site*". Inspections can be conducted from the ground and also using overflight by fixed or rotary wing aircraft. In addition to visual observations including photo and video, a number of measuring techniques may be used. For overflight the following equipment may be used: multispectral imagery, gamma spectroscopy and magnetic field mapping. In addition to visual observation and photographic and video recordings, the ground-based techniques may include radioactivity measurements like noble gas (xenon) surveys, passive seismological monitoring of possible aftershocks, active seismic surveys to look for underground anomalies, including cavities, magnetic and gravity field mapping and ground penetrating radar and electrical conductivity measurements. The last step is drilling into what is presumed to be the site of a nuclear explosion to obtain radioactive samples as final proof of a nuclear explosion. The protocol to the Treaty contains a number of details on how to conduct an on-site inspection.

4.4.4 Confidence-Building Measures

The confidence-building measures in the Treaty are related to chemical explosions and have two purposes: to prevent large chemical explosions from being misinterpreted as nuclear explosions and to use chemical explosions to calibrate the seismic component of the IMS. States undertake to cooperate with the Organization and provide the Technical Secretariat, on a voluntary basis, with notification of "*any chemical explosion using 300 tonnes or greater of TNT-equivalent blasting material detonated as a single explosion anywhere on its territory*". An underground explosion of 300 tonnes of high explosive, detonated as a single charge, was supposed to be detectable by the IMS. As the network, in many regions, might prove more capable than originally anticipated, considerably weaker explosions might also be detected.

The formulation "*detonated as a single explosion*" is intended to distinguish such an explosion from the ripple fired or interval blasts carried out as part of routine mining activities. Such a blast, which could have a large aggregate explosive energy, is composed of a number of smaller explosions detonated one after another. Such explosions generate weaker seismic signals. A summary of such ripple fired explosions, with an aggregate yield greater than 300 ton, should, on a voluntary basis, be reported as soon as possible after the Treaty has entered into force and at annual intervals thereafter.

4.5 National Implementation Measures

Article III of the Treaty deals briefly with national implementation measures. This article states that "*Each State Party shall, in accordance with its constitutional processes, take any necessary measures to implement its obligations under this Treaty*". It then continues that States should enforce any prohibitions in the Treaty on its territory or by any of its citizens. The final paragraph 4 in this article relates closely to the verification of compliance with the Treaty and it reads: "*In order to fulfil its obligations under the Treaty, each State Party shall designate or set up a National Authority and shall so inform the Organization upon entry into force of the Treaty for it. The National Authority shall serve as the national focal point for liaison with the Organization and with other States Parties*".

Article IV of the Treaty lays down a number of obligations of States Parties related to establishment and operation of the verification regime. To highlight a few: "*Each State Party undertakes in accordance with this Treaty, to cooperate, through its National Authority established pursuant to Article III, paragraph 4,*

with the Organization and other States Parties to facilitate the verification of
compliance with this Treaty by, inter alia:

(a) *Establishing the necessary facilities to participate in these verification mea-*
    sures and establishing the necessary communication;
(b) *Providing data obtained from national stations that are part of the Interna-*
    tional Monitoring System;
(c) *Participating, as appropriate, in a consultation and clarification process;*
(d) *Permitting the conduct of on-site inspections; and*
(e) *Participating, as appropriate, in confidence-building measures."*

This article also deals with possible improvements to the verification regime
and states that "*Each State Party undertakes to cooperate with the Organization*
and with other States Parties in the improvement of the verification regime, and in
the examination of the verification potential of additional monitoring technologies
such as electromagnetic pulse monitoring or satellite monitoring, with a view to
developing, when appropriate, specific measures to enhance the efficient and cost-
effective verification of this Treaty".

The article also promotes cooperation among States as follows: "*The States*
Parties undertake to promote cooperation among themselves to facilitate and
participate in the fullest possible exchange relating to technologies used in the
verification of this Treaty in order to enable all States Parties to strengthen their
national implementation of verification measures and to benefit from the applica-
tion of such technologies for peaceful purposes".

4.6 Entry into Force

The most controversial article of the CTBT is article XIV on entry into force,
which states: "*This Treaty shall enter into force 180 days after the date of*
deposit of instruments of ratification by all States listed in Annex 2 to this
Treaty, but in no case earlier than two years after its opening for signature".
Annex 2 lists 44 States that all had nuclear activities at the time of the
negotiations. By February 2008, more than eleven years after the Treaty was
opened for signature (Fig. 4.1), only 35 of these 44 States had ratified. The
nine States that prevent the Treaty from entering into force are China, Egypt,
India, Indonesia, Iran, Israel, North Korea, Pakistan and the USA. Despite
178 signatory States and 144 ratifications by February 2008, the Treaty is still
far from entry into force.

This entry into force requirement differs substantially from those in the
Chemical Weapons Convention (CWC) and the Biological Weapons Convention (BWC). As noted in Chapter 3, the CWC requires that 65 States ratify the
convention for entry into force and the BWC requires 22 ratifications, including
those of the three depositary States: the USA, the UK and the former USSR.

Fig. 4.1 President Clinton
signing the Comprehensive
Nuclear-Test-Ban Treaty
during a ceremony at the
UN Headquarters in New
York on 24 September 1996

The CWC entered into force on 29 April 1997, just over four years after it was
opened for signature on 13 January 1993 (OPCW 2007). The corresponding
time for the BWC was less than three years, from 10 April 1972 to 26 March
1975 (OPBW 2007).

4.7 Reflections

More than eleven years have passed since the Treaty was opened for signature
and work to implement the Treaty at the CTBTO Preparatory Commission
commenced. Our intent here is to provide some reflections on the Treaty
provisions in light of experiences during these years.

4.7.1 Politically Significant

The very fact that the CTBT was finally signed was an essential political
achievement. The different elements of the Treaty, in particular the

comprehensive verification arrangements, are all important, but the crucial thing was to get agreement on a treaty. Discussions and negotiations had been going on for some 40 years and the lack of a test ban treaty was a show stopper for other initiatives. It also had a most harmful effect on existing treaties, such as the NPT, where the lack of a CTBT seriously affected the review conferences.

Since it has not gone into force after more than eleven years, the CTBT is not making the full contribution it could to international efforts in the fields of non-proliferation and disarmament. Although nobody expects that any State, having signed the CTBT, will resume testing, it is politically most essential that the Treaty will enter into force soon.

The ambition to involve all key States from the beginning, using a unique entry into force clause, is a most respectable ambition but it has to be weighed against a long or very long delay in the entry into force (EIF). With a long non-EIF period it could prove difficult to sustain both the political interest in the Treaty and the support for the preparatory work. There could be fading interest in financing and supporting a large verification system and it might also be difficult to attract qualified experts to join the work of the CTBTO Preparatory Commission. The CTBT has proved that a great risk is involved when you create a mechanism that makes it possible for a small number of States, or just a single State, to hold a treaty hostage. Given the present experience it might have been more attractive to follow a conventional entry into force procedure that would have brought the CTBT into force long ago. This might also have created more powerful pressure on States that so far have not signed or ratified the Treaty.

4.7.2 The Executive Council

The Executive Council is an important component of the CTBTO, as, among other tasks, it has a crucial role in rejecting or approving requests for on-site inspections. The composition of the Executive Council was an important issue during the negotiations. The membership of the Executive Council is elected by regional groups. In the CTBT a new grouping was introduced, a Middle East and South Asia group including a number of Arab States and Israel; this group carries the heavy burden of the political situation in the Middle East. The argument was that this was a purely geographic grouping. Correct as this might be, the Treaty is very much about politics and that is where the problems arise. In the CTBTO Preparatory Commission it has proved impossible to even conduct a meeting within this regional group, illustrating the influence of the actual political situation. To some extent this is inevitable but it is essential not to create arrangements or procedures that may unnecessarily amplify such influence. To bring Israel into the North America and Western

Europe group, with which Israel already has a number of connections, might be geographically incorrect but would have removed a serious political problem in the CTBT.

4.7.3 The Role of CTBTO and States Parties

The CTBT aims at giving all States an equal and fair possibility to monitor compliance with the Treaty by providing high quality information to all States. It also provides for most intrusive on-site inspections. The CTBTO has a quite different and more limited mandate when it comes to verifying compliance with the CTBT compared, for example, to the IAEA in relation to the NPT. CTBTO manages the IMS and provides States all data collected and a number of products, also custom designed. Annex 2 to the Protocol of the Treaty identifies a number of parameters that might be used to further characterize the observed events. The parameters given are tentative and the dramatic development in information analysis since the Treaty was signed should make it possible to further develop the methods of characterizing or screening observed events. These products are of crucial importance for many States that do not have the capability to do their own routine processing of the large volume of data routinely collected by the monitoring network. The final analysis and the political assessments of observed events rest with States Parties. The assessment of whether or not an observed event needs further clarification is a political one. It will be based on an interpretation of the observations from the event and an assessment of the political and technical situation in the State in which the event occurred.

Our interaction with delegates at the CTBTO Preparatory Commission has allowed us to note that this national responsibility to make the assessment of the information provided is not yet fully realized by all. Many seem to have the misconception that the Organization will do the job and tell them when a suspicious event has been observed. States in many parts of the world do not have the necessary technical facilities nor the expertise to do the necessary interpretation of the data provided by the Organization. Among the developed States, too, the resources to interpret the information provided have been cut. Many States had large research facilities and research groups to support the development and testing of the verification system prior to and during the negotiations, but many of these groups have declined or disappeared. Today only a few States have the ability to do full, independent data interpretation and make an assessment as to compliance with the Treaty. If the assessment on compliance is not to stay in the hands of a handful of States, more States have to increase their efforts and resources in anticipation of the Treaty's entry into force. One way that has been informally discussed is to create regional centers through which a particular region combines resources to effect a joint interpretation. Such regional centers could be established in Europe, Africa, Latin America and Asia. We return to this idea in Chapter 8.

4.7.4 An Unprecedented Verification Regime

The international verification measures in the CTBT are unprecedented. The CTBT proves that it is possible to reach agreement on a highly technical, complex verification system with a global reach and an intrusive on-site inspection regime. This in itself is a significant confidence building measure. As we discuss in detail later in the book, the implementation of these measures has progressed during the long period of the CTBTO Preparatory Commission. At a slower pace and at a higher cost than originally anticipated, the last eleven years have nevertheless proved that it is possible to physically establish and provisionally operate an extensive global verification system. This is an important experience that might encourage similar efforts for other applications.

The Treaty provides little guidance on the technical specifications of the stations in the IMS; this was left to the CTBTO Preparatory Commission to decide, which proved to be a wise decision. Thorough work was carried out by the CTBTO Preparatory Commission on the technical specifications for the different kinds of stations and how they should be certified to meet those specifications. The provision in the Treaty giving States the possibility to build and finance a station and reduce its assessment by a corresponding amount was created to initially speed up the implementation of the network. The downside of this well-intended idea was that a variety of station equipment was introduced, which may make operation and maintenance more costly in the long run.

The Treaty also contains provisions for a most intrusive on-site inspection regime, giving access to a large area and using a large number of advanced technical tools. This too is a good example for future treaties. The protocol to the Treaty contains a lot of detail on methods and procedures to conduct an on-site inspection. Many of these procedures have been revisited during more than ten years of detailed elaborations within the CTBTO Preparatory Commission on an Operational Manual for on-site inspections. Given the limited shared knowledge about how to conduct on-site inspections at the time of the negotiations, it might have been wise to limit the amount of detail in that protocol.

The Treaty is based on science and technology available at the time of the negotiations, which is quite natural. Developments have been and still are dramatic in many scientific and technological areas crucial to the CTBT, including communications, hardware and software, and sensor technology. How to make use of these important developments in science and technology to improve the quality of the products provided by the Organization and to increase the cost-efficiency of the system is a most essential issue. To have an efficient procedure for reviewing and updating the technical systems and analysis methods and procedures on a regular basis is essential to the maintenance of a viable system.

This is not only a question of updating technologies that are part of the Treaty but also to benefit from new developments that were simply not generally available when the Treaty was negotiated. GPS and similar navigation systems are today in common use around the world. The same is true of high resolution satellite photos that today are only a few keystrokes away on our computers. It is essential to be able to make full use of such new technologies and methods.

4.7.5 Verification, Too Little or Too Much?

CTBT has extensive and elaborate verification measures, beyond any other international treaty. How much verification is needed? What constitutes adequate verification? During the negotiations the British Ambassador once came up with the following definition: "*Adequate verification is verification that satisfies all concerned*". The verification measures in the Treaty were considered adequate at the time of signature as they satisfied all concerned. Is it still adequate or is it maybe even too much? Are there ways to make verification more cost effective? It is certainly too early to make a comprehensive assessment of the capabilities of the different elements of the CTBT verification system. With the experience gained so far and at some distance from the negotiations, it might still be of interest to make a few comments.

The seismic component is a most basic element of the IMS, since it is the primary way to monitor underground events. The primary network of the two-tiered seismic system comprising 50 stations was well developed and tested at the time of the negotiations and has proved to be a most capable system. The auxiliary network of 120 stations carries with it both great potential and a problem. It was originally intended as a cheap way of benefiting the analysis using data, when needed, from existing stations operating for other purposes. The operational costs of these stations are to be covered by the hosting States. The auxiliary network has now developed into a high quality network where stations have been upgraded to the same technical standards as those of the primary stations. Data from the auxiliary stations should be provided on request only. This is partly a political decision and partly a reflection of the ability and cost of global communications in the mid-1990s. Today it has proved more efficient and economic to collect all data, including those from these stations, at the IDC and then select what you want to use. Even during the work of the CTBTO Preparatory Commission, questions have been raised on the funding of the auxiliary stations, as have questions on a more extensive use of these data, given that they can be made readily available, if decided. So we are facing two questions. Do we really need the auxiliary network given the capability of the primary network? If we keep it, should we not then make full use of the data it produces? The inclusion of auxiliary seismic stations in the event detection stage would lower the event detection threshold of the seismic network. This is discussed further in Chapter 6.

It might be interesting to address the contribution from the hydroacoustic underwater installations, which have proved quite complex and expensive. Are they providing crucial monitoring information or could they be replaced by simpler and considerably cheaper installations on shore? Or is the seismic system adequate to cover the oceans, too?

A comprehensive infrasound network is part of the verification regime to primarily monitor explosions in the atmosphere. Infrasound stations were operated in several States at the time of nuclear atmospheric testing, but were then dismantled. This is the first time ever that such a global network has been established and, obviously, experience with its performance is lacking. We shall soon get the results from the first tests of the network and it will be interesting to see how much it adds to the overall monitoring capability. Is it a cost-effective contribution, given the unexpectedly high cost of the system, or would satellite data that are generally available today be more cost-effective? Satellite data are today recognized in the Treaty as a national technical means.

The radionuclide networks for detecting radioactive particles and noble gases are also natural elements in the verification system, as observing radio-active products is the only way to prove that an event is nuclear. There is a fair amount of national experience with the monitoring of radioactive particles, but this is the first time that data will be available from an extensive global network. The technology for detecting radioactive noble gases is quite new and much of it has been developed and successfully tested during the CTBTO Preparatory Commission phase. The global capabilities of these systems still have to be demonstrated, in particular the ability to track the observed radioactivity back to the source.

Our assessments of the performance and capabilities of the various elements of the CTBT verification regime, in their current state of implementation under the auspices of the CTBTO Preparatory Commission, are provided in Chapter 7.

Chapter 5
The Birth of An Organization

In the previous chapters we have described the CTBT and the negotiations that created the Treaty, a process that culminated with the opening for signature of the CTBT on 24 September 1996. During the final stages of the CTBT negotiations in the CD, the negotiators faced the additional challenge of laying the foundation for the organization that on behalf of all States Signatories would be entrusted with the task of preparing for the entry into force of the CTBT. The negotiations of the "Text on the Establishment of a Preparatory Commission for the Comprehensive Nuclear-Test-Ban Treaty Organization" (briefly, the CTBTO Preparatory Commission, or, as also used in this book, the PrepCom) were successfully concluded in the CD. This text was then adopted in a resolution passed at a meeting of States Signatories in New York on 19 November, 1996, one day prior to convening the first session of this new organization.

In this chapter we give a fairly detailed account of events and processes surrounding the birth and infancy of the PrepCom, in an attempt to address some key questions: Which were the issues that dominated the agenda of the organization's first sessions, in which the foundation was laid for efforts to prepare for entry into force of the CTBT? What were the challenges during the first few years of the PrepCom, and how was the enthusiasm and eagerness of all those who had been looking forward for such a long time to contribute to the work of the PrepCom channelled in the right direction? And were there early signs of issues that have proved difficult for the PrepCom over a long time and that are perhaps still with the organization today? First, though, let us take a closer look at the text of the resolution that established the PrepCom. The full text of this resolution can be found on PrepCom's home page (CTBTO Preparatory Commission 2007a).

5.1 The Mandate of the CTBTO Preparatory Commission

According to the text of the resolution establishing the PrepCom, its purpose is to carry out "*the necessary preparations for the effective implementation of the Comprehensive Nuclear-Test-Ban Treaty, and for preparing for the first session of*

O. Dahlman et al., *Nuclear Test Ban*, DOI 10.1007/978-1-4020-6885-0_5,   99
© Springer Science+Business Media B.V. 2009

the Conference of the States Parties to that Treaty". PrepCom is mandated to *"undertake all necessary preparations to ensure the operationalization of the Treaty's verification regime at entry into force"*. This includes the establishment, testing and provisional operation of the International Monitoring System and the International Data Center. It also includes making the necessary preparations for the conduct of on-site inspections. On the political side PrepCom shall *"follow the ratification process and, if requested by States Signatories, provide them with legal and technical advice about the Treaty in order to facilitate its ratification process"*. PrepCom is thus given a strong mandate and an onerous task to establish and test the Treaty's verification regime to ensure that it is operational at entry into force. On the other hand, PrepCom is given only a vague mandate to support the ratification process, if requested by States Signatories to do so.

Thus, the PrepCom is the organization that carries the CTBT torch until the closing of the first session of the Conference of the States Parties, which shall be convened no later than 30 days after the Treaty's entry into force. The PrepCom is composed of all States that sign the Treaty. The text of the resolution establishing the PrepCom stipulates that the costs of the PrepCom and its activities shall be met annually by all States Signatories, in accordance with the United Nations scale of assessment, adjusted to take into account differences in membership between the United Nations and the PrepCom. States Signatories that have not discharged in full their financial obligations to the PrepCom within a year of receipt of request for payment shall have no vote in the PrepCom plenary body. All decisions of the PrepCom plenary body should be taken by consensus, but procedures are in place to make decisions based on voting, if consensus cannot be found. The PrepCom elects its chairperson and other officers, adopts its rules of procedure, decides on its meeting schedule and establishes committees as deemed useful. It appoints its Executive Secretary and establishes a Provisional Technical Secretariat (PTS) to assist the PrepCom in its activity and to exercise such functions as the PrepCom may determine. Prepcom establishes the annual work program and budget for the PTS and oversees its activities. PrepCom is also mandated to establish administrative and financial regulations in respect of its own expenditure and accounts.

The text of the resolution establishing the PrepCom gives, as we have seen above, some general guidance on the technical work needed in order to establish the CTBT verification regime in preparation for entry into force. An appendix to this text provides additional details in the form of an indicative list that *"is illustrative of the verification-related tasks the Preparatory Commission might need to undertake in implementing the relevant provisions of the Treaty and of the resolution establishing the Commission"*. Many of these tasks are related to the development of the Operational Manuals called for in the Treaty for the various IMS technologies, the IDC and the OSI. The PrepCom is mandated to present a report on the operational readiness of the verification regime, together with any relevant recommendations, to the initial session of the Conference of the States Parties, following the Treaty's entry into force. All PrepCom's work is provisional, in the sense that everything related to activities of the future CTBT Organization (CTBTO) after entry into force is up for review, possible revision and approval by this conference. It is envisaged, though, that thorough and systematic work during the PrepCom phase, conducted with the consent of States Signatories, will ensure a smooth transition.

5.2   Preparations for the First Session of the CTBTO Preparatory Commission

During September to November 1996, interested delegations, "Friends of the CTBT", met in informal, open-ended discussions in Geneva and New York to prepare for the first plenary session of the PrepCom, under the coordination of Canada. Their work, guided by the existing, but not yet adopted, text on the establishment of the PrepCom, was aimed at preparing decisions related to institution-building needed to enable the PrepCom to function properly from the outset. Decisions were prepared for the appointment of an Executive Secretary, administrative and financial rules and regulations, a budget and work program for the first few months, the structure of the PTS and the creation of subsidiary bodies to the PrepCom plenary body. The initial PrepCom session also had to consider a host country agreement for the PrepCom. The text on the establishment of the PrepCom states that its seat shall be at the seat of the future CTBTO and Austria had already made an offer to host the CTBTO in Vienna.

Agreement was readily found in these discussions to propose to the first session of the PrepCom that two working groups should be set up: Working Group A for budgetary and administrative matters, and Working Group B for verification. The PrepCom text is not explicit on such subsidiary bodies; it only states that the PrepCom shall "*establish such committees as it deems useful*". Proposals for creation of these subsidiary bodies were clearly in line with experience gained during the CTBT negotiations, in the sense that States had found it useful to conduct detailed discussions in dedicated, subsidiary bodies. Working Group B for verification would also build upon experience from CD's Group of Scientific Experts (GSE).

Discussions on the structure of the PTS encountered difficulties. Various organograms for the PTS were presented but none were agreed during this preparatory phase. One difficulty was that many delegations held the view that there should be one director level position for each of the six regional groups defined in the Treaty. On top of this, it was not clear to all whether the Executive Secretary position should be counted in this regard. Moreover, some delegations favored the inclusion in the PTS structure of a director level post in the office of the Executive Secretary for coordination and top level management of the prospective IMS, IDC and OSI divisions. A similar proposal was to create a position for a Deputy Executive Secretary. Things were not made any easier when individual countries started to launch their candidates for top level positions, even during the discussions on the structure of the PTS. It was eventually left to further consultations during the first session of the PrepCom to finalize a structure for the PTS.

5.3 The First Session of the CTBTO Preparatory Commission

As we have noted, the States Signatories gave birth to the PrepCom in their meeting on 19 November, 1996 in New York, through the adoption of a resolution containing the PrepCom text. The Signatory States then met at the same place during 20–22 November for the first session of this new organization. Ambassador Jacob S. Selebi of the Republic of South Africa was elected chairman of the PrepCom for its first session.

It soon became clear that the ambitions for this session, as reflected in its agenda, could not be met. The structure of the PTS and allocation of positions to regional groups and individual countries dominated the discussions. It was not possible to find consensus on these matters, in spite of tireless efforts in late-hour consultations. This implied that the PTS could not be established at this meeting. This lack of agreement also spilled over into other matters that were up for decision: there was no appointment of an Executive Secretary, although Ambassador Wolfgang Hoffmann of Germany was the only candidate and enjoyed broad support; no one was given the authority to sign the host country agreement with Austria although the agreement as such was approved; the two working groups were not established; and the proposed work program and budget for the first four months were not approved. On the other hand, provisional rules of procedure of the PrepCom were adopted and so were provisional financial regulations as well as staff regulations and rules.

It was decided to reconvene the first session of the PrepCom in Geneva in March 1997. The chairman was given the mandate to engage delegations in consultations in the interim period to address the remaining problems. In his closing remark on 22 November 2006, the chairman commented that the four months before the resumption of the first session would be needed to find solutions to the outstanding agenda items.

Consultations in preparation for the resumed first session of the PrepCom started in Geneva in December 1996 and continued through February 1997. Ambassador Selebi named Ola Dahlman, Sweden, his Friend of the Chair for verification, with the task of supporting the preparation for the resumed first session of the PrepCom within that area. Ambassador Mark Moher, Canada, was appointed Friend of the Chair for developing a draft budget for the first year of the PrepCom. The consultations conducted by these two Friends were successful and facilitated decisions subsequently made at the resumed session in March. In addition, consultations were held that finally resolved the contentious issues related to the PTS structure and the filling of its top level posts.

The verification-related consultations concentrated on the work program for the PTS and also a work program for Working Group B, in anticipation of its establishment at the resumed session. The future organization of work in WGB was also addressed. Experts in all IMS technologies and in aspects of IDC establishment participated in these consultations. Based on the status of existing IMS-designated stations that were participating in the GSETT-3 experiment, a work program to initiate the establishment of IMS was developed. These discussions showed that experts in the four technologies were thinking in terms of a timeframe of only some three years to essentially complete the IMS network, under the assumption of adequate budgets and the absence of any legal and administrative obstacles. Even though the attitude might have been colored by the expectations at that time of an early entry into force of the Treaty, these experts clearly saw it as possible, from a technical point of view, to complete the establishment of the IMS within a few years. These consultations also showed early indications of some themes and issues that would later prove to be problematic for PrepCom, such as the mode of operation of the IMS/IDC during the PrepCom phase, the development of the noble gas component of the IMS radionuclide network, and the ability of experts from developing countries to participate fully in technical discussions conducted in the PrepCom.

The PrepCom then met during 3–7 March 1997 in Geneva and successfully completed its first session. The PrepCom named Ambassador Wolfgang Hoffmann of Germany as its Executive Secretary for an initial term of two years. It further established Working Group A for budgetary and administrative matters and appointed Ambassador Tibor Tóth as its chairman, it established Working Group B dealing with verification issues and appointed Ola Dahlman of Sweden as its chairman, and it approved work programs for these two working groups for the remainder of 1997. The PrepCom also established the PTS in Vienna with an agreed-upon structure containing five divisions (Administration, Legal and External Relations, IMS, IDC and On-Site Inspection) and appointed its five directors, and adopted its budget of $27.5 million for 1996 (for expenses already incurred) and 1997. This budget included $8.2 million for the establishment of IMS stations, $4.5 million for initial investments for the IDC, $7.1 million for staff positions, which would provide for the filling of 116 posts by the end of 1997, and $7.7 million for conference services, PTS operations and a contingency reserve.

With this outcome of its first session, the PrepCom was in a good position to embark on its challenge to establish the CTBT verification regime. The PTS took up work in its office facilities in the Vienna International Centre (Fig. 5.1) on 17 March 1997 with a staff of 10 people. The number of staff members was gradually growing; see Figs. 5.2 and 5.3. It turned out that the work program and budget for the PTS for its first year was overly optimistic, as only $13.8 million or 50% of the allocated budget had been spent by the end of 1997.

Since this first session, the PrepCom has been organized with its Policy Making Organs and its executive arm, the PTS. The Executive Secretary of the PrepCom heads the PTS, as its chief executive officer. The superior Policy

Fig. 5.1 View of the Vienna
International Centre,
Austria. The Provisional
Technical Secretariat shares
these office buildings with
other international
organizations such as IAEA
and UNIDO

Making Organ, the PrepCom plenary body, meets in quite short sessions open
to all States Signatories, initially three times a year. The chairmanship of the
PrepCom, initially for a duration of half a year and later extended to one year,
rotates between the regional groups. PrepCom has, as we have seen, two
subsidiary Working Groups. Both Working Groups are open to all States

Fig. 5.2 Employees of the Provisional Technical Secretariat in 1997

Fig. 5.3 Employees of the Provisional Technical Secretariat in 2007

Signatories and most of the work in PrepCom on elaborating individual issues takes place in these Working Groups. The Working Groups provide reports to the PrepCom containing recommendations to be acted upon via decisions in PrepCom plenary sessions. The chairs of Working Groups A and B were initially appointed without time limitations to their tenures. Initially, the Working Groups met during, or in conjunction with, PrepCom sessions and also in the interim between PrepCom sessions, but since 1998 the subsidiary bodies have only met between PrepCom sessions, at a frequency of two or three sessions per year. PrepCom has also created an Advisory Group that advises the PrepCom primarily on financial and budgetary issues. The Advisory Group initially consisted of twelve appointed experts. Changes during recent years to terms of appointment and duration of sessions are summarized in Chapter 9.

The issue of how to create appropriate structures to regularly provide the PrepCom with expert and integrated technical advice on all aspects of the establishment of the verification regime was widely addressed and discussed during these early stages of the PrepCom. The primary mechanism adopted for this is reflected in the method of work of Working Group B, which relies on task leaders who organize the work in task leader sessions, within the framework and meeting time of the Working Group. The findings and conclusions from these task leader sessions, which consume most of the time allotted to Working Group B, are presented in the plenary sessions of the Working Group for incorporation into its report to the PrepCom. In some instances, these task leaders have set up open-ended expert groups to consider specific technical issues, and advice emerging from such groups has been channelled back to the working group via the task leader. Working Group A has conducted most of its work in plenary sessions, but has also relied on assistance from Friends of the Chair and "focal points" appointed to deal with specific issues. Experience over the years from an organizational perspective is further discussed in Chapter 9.

5.4 The Initial Enthusiasm

With the positive outcome of the first session of the PrepCom in terms of a working structure and a budget, efforts could start right away both in the PTS and in the Policy Making Organs. There was a clear sense of urgency in most quarters, and a widely shared understanding that it would be possible to establish the verification regime within a time frame of three to four years, provided adequate resources were made available by the States Signatories. On the technical side, it was necessary to immediately start activities in Working Group B to address the major task of establishing the IMS, the IDC and the new Global Communications Infrastructure to provide for transmission of data from IMS stations to the IDC, as well as products from the IDC to National Data Centers. An example of such early activities was the development of technical requirements and specifications for IMS stations in all four technologies. Such specifications were agreed in the Working Group already in April 1997, and adopted by the PrepCom at its May 1997 session. Following this, procurement and installation of IMS station equipment could start, and existing stations could be upgraded to meet agreed standards. Similarly, specifications were developed and adopted for IDC hardware and software, and a seven-phase commissioning plan for the IDC was in place at an early date. Specifications for the communications infrastructure and a plan for its procurement were other tasks that were addressed early on. All these issues were considered by States Signatories as matters of policy, and thus needed to be dealt with in the Policy Making Organs before the PTS could start executing its work. Work also progressed rapidly in Working Group A, and within a short time necessary rules and regulations for the proper functioning of the Policy Making Organs and the PTS were developed, agreed and adopted by the PrepCom.

   In contrast to the situation during the negotiations, in which the atmosphere naturally was tainted by conflicting national positions and the need to hammer out compromises on complicated matters, a spirit of cooperation prevailed during the early stages of the life of the PrepCom. This new era represented a long awaited opportunity to leave confrontational politics behind for a while and concentrate fully on the task of establishing the verification system. For this task, regular, and we may say generous, budgets were now available to pay for program costs, including salaries of PTS staff. This was also in contrast to the funding of technical preparatory work like the GSETT-3 experiment, which relied on voluntary contributions by participating countries.

Numerous people had a strong wish to contribute to the work of the PrepCom. Many individuals who joined the PTS had considerable experience from the preparatory work, either as delegates to the GSE or as members of national delegations to the negotiations, or both. Others chose to continue work at their home institutions but got involved in the work as national delegates to Working Group B. Yet others joined their permanent missions in Vienna and participated from that base in the Policy Making Organs of the PrepCom. During the early stages, when the PTS was in its infancy and relatively few positions had been filled, the Policy Making Organs supported the PTS by conducting work of a nature that was later carried out by the PTS. This situation, coupled with the enthusiasm of those early years, led to a kind of "family sense" from which the work benefited greatly. Many people who had been involved with the CTBT issue for a long period, some for more than 30 years, thus wanted to contribute, together with others, to finally see the CTBT coming to fruition and enter into force.

As we have noted, most of the work in Working Group B is led, organized, facilitated and coordinated by task leaders, on behalf of the chairman of the Working Group. Task leaders are chosen in view of their technical expertise. Their assignment is to ensure that the best available technical advice and guidance are provided to the Working Group and the PTS, reflecting deliberations among States Signatories. Task leaders thus work actively with delegations to achieve agreement on all issues in their area of responsibility. The first group of task leaders started work immediately following the first meeting of the Working Group in March 1997 and Working Group B has relied on dedicated efforts by task leaders to this date. Task leaders are members of their respective delegations to the Working Group, and are supported by their own countries to fulfil their functions on behalf of all States Signatories. Given the enthusiasm and the availability of highly competent individuals in many States Signatories, it was initially easy to find people who were able and willing to take on the responsibility of being task leaders. Likewise, their countries were in general prepared to provide the necessary support, as a voluntary contribution to the PrepCom process. Over the years, these contributions have been most substantial, considering all the work performed by task leaders. Annex 2 to this book provides the names of task leaders of Working Group B, their tasks, countries of origin and periods of service. This annex also provides names and tasks of focal points and friends of the chair of Working Group A together with the Executive Secretaries and the PrepCom chairs.

The delegates to the PrepCom sessions and to the working groups have two basic tasks; to promote the interests of their own country and to contribute to moving the PrepCom process forward. We have seen over the years that delegates have generally engaged constructively with each other in the consensus building processes. Delegations to Working Group B have also contributed to the work through presentation of national papers on various technical and scientific topics. Such contributions were particularly important in the early years for discussion and resolution of many complicated technical issues, at a time when there were few PTS employees to support the discussions.

5.5 The Early Challenges

It was a challenge in the early days to channel all the enthusiasm in the right direction and to try to do things in the right order at all levels. This involved holding back in some areas to avoid rushing into decisions, even under great pressure, political and otherwise, to show good progress. An example here was the self-imposed restraint not to embark on station installations before technical specifications had been fully addressed and agreed. Another important consideration in the early stages was to make sure that work started early on issues with long lead times; an example was the early concentration on the conduct of site surveys to enable station installations to proceed without unnecessary delays. More generally, it was an overarching objective, subscribed to by all stakeholders in the process, to "*increase the preparedness and flexibility to finalise the verification system so as to meet future needs*". This is carefully negotiated language for the idea that the implementation of the verification system should be planned in such a way that the time needed for its completion should be as short as possible, when the Treaty's entry into force would be within reach and appropriate resources would be available for the final ramp-up activities. This is a great challenge when building a complex technical system with a global reach in a political environment. The strategic planning needed is complicated, as the establishment of monitoring stations also depends on the sometimes unpredictable decisions and actions by States. Nor do the Policy Making Organs provide an environment conducive to long term planning of complex technical systems. As we elaborate further later, the PTS initially attracted most competent scientists, but relatively few people with managerial or strategic planning experience. Our experience from the activities in these first years is that a lot was done opportunistically rather than according to a well staged plan. Given the circumstances, this was probably not as bad as it may sound.

Some early decisions of the Policy Making Organs were made on the premise that the verification system with its IMS and IDC components would be in place within three to four years, and that the operational requirements during the PrepCom years would fairly closely match those that would apply after the Treaty's entry into force. For example, requirements for the new communications system and guidelines for its procurement were developed rather quickly, leading to an early signature of a contract with a communications provider, which started the roll-out of the communications system in August 1998 immediately following the signature of the contract. As we shall see, it soon became evident that the timeline for the completion of the verification system would change and that key operational requirements during the PrepCom phase, like those related to station uptimes, would be relaxed. Fortunately, the PTS succeeded at a later stage in renegotiating the communications contract to bring the implementation in phase with changed circumstances resulting from events at the political level. One of the lessons learned here is to make sure that decisions with effects far into the future are sufficiently flexible to allow adaptation to moving political targets like a treaty's entry into force.

Installation of IMS stations around the world requires a legal framework in the form of arrangements or agreements between the PrepCom and each of the 89 countries hosting these stations. Ideally, such facility agreements, regulating legal, financial, taxation and a range of other issues related to IMS station installation and subsequent operation, shall be signed and shall have entered into force before activities are started in the host countries. Experience has shown that it takes a very long time to get these agreements in place, and it has taken more than 10 years to sign 37 agreements. So instead of waiting for such agreements, much of the installation work has proceeded based on a simplified exchange of letters with the hosting countries, containing text that provisionally regulates the relations, pending future agreements. At the time of writing (late 2007) it has not yet been possible to reach any sort of agreement with five countries to proceed with IMS station installation. Over the years, the PTS has thus to a certain extent been obliged to shape its IMS installation program in accordance with opportunities provided by such agreements with host countries. A lesson learned here is that clearly stated political will and assurances of full cooperation in the establishment of IMS stations is not enough; such statements must be followed up by agreements on concrete preparations in the various host countries.

Initially much of the effort and interest focused on the build-up of the IMS and the IDC. OSI ended up on the back burner, not as a result of any deliberate decisions but rather because of lack of initiatives both in Working Group B and in the PTS. It is quite clear that the implementation of the OSI regime was lagging behind and it would have required a most concerted effort to make it operational within a period of 3–4 years. This is especially true of the OSI Operational Manual, where essentially no progress was made during the first years. The main reason for this was political. A lot of political tension from the CTBT negotiations was still a heavy backpack for several of the delegations during the elaboration of OSI procedures.

The wellbeing and success of an organization like PrepCom depends strongly on the sense of ownership that can be developed among its stakeholders. This sense of ownership must apply to the consensus building processes itself, as well as to every decision made by the Policy Making Organs. With all the enthusiasm present at the outset and the clearly expressed will of working towards a common goal, there were high hopes in this regard. And it is fair to say that a culture of ownership did develop. Not least, a robust and resilient decision making process developed during these early years in the PMOs that gave all participants an opportunity to have their say, and there was a widespread willingness to accommodate concerns of others to facilitate consensus. The mood in the Policy Making Organs has, however, become more confrontational during the last few years, as we shall see in Chapter 9.

Much attention was devoted at an early stage to defining the proper roles for the PTS on one hand and the Policy Making Organs on the other, and to developing mechanisms for constructive interaction between them. As we have seen, the roles were intentionally a little mixed in the beginning, when much work on technical implementation detail was done under the auspices of Working Group B, since the PTS was still in a build-up phase and had not yet acquired necessary staff. For example, the verification-related budget proposals for 1998 were developed by Working Group B. Efforts to develop the verification-related parts of the budget proposal for 1999 were shared between the PTS and Working Group B. For the budgets for the following years, an annual cycle has been followed, with clearly established roles based on budget proposals by the PTS and review by the Policy Making Organs, and ultimately a decision by the last PrepCom session of the year in November.

With respect to the roles of the PTS and the Policy Making Organs, a useful concept is for the Policy Making Organs to concentrate on *what* should be done, leaving it to the PTS to decide *how* to execute the *what*. By and large this understanding of the roles was already in place in 1997 and still remains the basis for the interaction between the PTS and the Policy Making Organs. At times, though, individual delegations or groups of delegations challenge this concept by suggesting guidance to the PTS that is perceived by the PTS as micromanagement. It is incumbent on the leaderships of the Policy Making Organs to try to balance guidance and requests directed to the PTS in such a way that the executive role of the PTS is not compromised. Experience has shown that controversies arising from lack of common understanding of roles do appear from time to time, but also that discussions of these roles tend to foster renewed and broader understanding of the way the interaction between the PTS and the Policy Making Organs should work.

Small countries, several developing countries in particular, raised concern early on about their representation and influence in the Policy Making Organs. Most specifically, the concern was about their possibility to take an active part in the technical discussions in Working Group B. In many cases, such countries did possess technical experts with relevant knowledge, but could not fund travel expenses for them to participate in sessions in Vienna. One undertaking that partly remedied this situation was the Experts Communication System, a website that posts documentation and supports technical discussions intersessionally for PrepCom participants. This tool, proposed by New Zealand and put in place by the PTS, is useful in particular for experts who are unable to attend the sessions in Vienna, although it does not allow delegates to participate online in those sessions, as originally proposed. More recent attempts to increase the insight into PrepCom activities have included a mechanism for online video streaming of sessions of the Policy Making Organs, enabling interested parties to follow the meetings from afar. Nevertheless, even today, the only way of ensuring influence is to be present at the sessions of the Policy Making Organs.

The consensus building processes in the Policy Making Organs always benefit from a geographically well balanced participation. The development of ownership of PrepCom decisions by less resourceful States Signatories is crucial to the universality of the PrepCom process and the development of its verification regime. One measure for fostering wide participation in the technical discussions conducted in Working Group B is to ensure a wide geographical representation in the group of task leaders, and also to include experts from developing countries in this group. Much effort has been expended over the years to engage task leaders from all around the world and it has usually not been difficult to identify excellent potential candidates in developing countries. The challenge has been for these countries to be able to sustain national funding over a longer period to keep such task leaders involved in the process. Over the years, several task leaders from developing countries have left their posts due to lack of resources, and this has yielded a task leader group for which a better geographical balance would have been desirable.

We have described above the first enthusiasm, the high expectations, the initial sense of urgency and the widespread belief that it would be possible to establish the verification regime within a matter of three to four years. When did all of this change, and why? The first signs of a different attitude emerged as early as the budget discussions in 1997, when delegations started to realize that costs would be higher than many had anticipated and several countries were concerned about their ability to pay their own share of the budget. This was expressed, after some time, and on behalf of several delegations, in terms of a requirement that the pace of establishing the IMS should be adapted to the prospects of entry into force of the Treaty. These prospects of entry into force were clearly affected by the US Senate decision in October 1999 *not* to ratify the Treaty. This event in the political arena was sufficiently significant in its impact that one could say that the first era in the life of the PrepCom, characterized by widespread enthusiasm and a sense of urgency, ended. At this juncture, a new era started.

Chapter 6
Establishing the Verification Regime

6.1 A Complex Monitoring System in a Political Environment

The CTBT verification regime is designed to provide confidence that nuclear explosions can be detected and identified as such anywhere and at any time. The monitoring component of this regime therefore needs to be global in its coverage and on the alert at all times to pick up signs of possible non-compliance with the Treaty. The system, now nearing its completion, is the most comprehensive technical system ever established to verify compliance with a treaty in the multilateral arena, and at the same time among the more ambitious undertakings ever to monitor all environments of the earth. The monitoring component of the regime is illustrated schematically in Fig. 6.1, which shows sensors of the International Monitoring System (IMS) providing data via the Global Communications Infrastructure to the International Data Center (IDC) of the Provisional Technical Secretariat (PTS) of the CTBTO Preparatory Commission, which in turn provides analysis results and, if requested, also raw data to National Authorities for their use.

In previous chapters we have described a number of technologies that have been considered over the years for possible inclusion in monitoring systems, and we have also seen that the CTBT negotiators settled on four such technologies: seismic, hydroacoustic, infrasound and radionuclide monitoring. The IMS, defined in the Treaty and comprising altogether 321 monitoring stations, is shown in Fig. 6.2. Radionuclide laboratories at 16 locations around the world (also shown in Fig. 6.2 and listed in Annex 1 to this book) support radionuclide monitoring stations by re-analysing samples of particular significance from these stations. As the figure shows, the distribution of monitoring stations is relatively even over the land masses of the world, and many small islands are hosting stations to provide a global balance. To exploit synergies between different technologies and reduce operational costs, stations of the various networks are sometimes co-located. Many stations are established at remote locations to minimize the impact from man-made disturbances. The remoteness of these locations often presents logistical challenges in the form of lack of infrastructure, such as access roads and external sources of electricity.

O. Dahlman et al., *Nuclear Test Ban*, DOI 10.1007/978-1-4020-6885-0_6,
© Springer Science+Business Media B.V. 2009

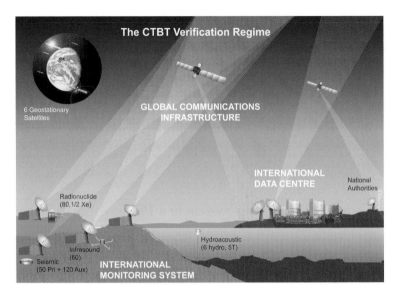

Fig. 6.1 Schematic view of the monitoring component of the CTBT verification regime. The figure illustrates sensors of the International Monitoring System providing data via the Global Communications Infrastructure to the International Data Center in Vienna, which in turn provides analysis results to National Authorities

The primary seismic network of 50 stations is a key element of the IMS, as seismic signals are likely to be the only or at least the initial sign of underground explosions. The continuous transmission of data in real time from this network to the IDC provides for the detection and initial location of seismic events worldwide. As many as 33 of the 50 stations of this network will be arrays. As we have seen in Chapter 2, a seismic array is composed of a number of sensors, arranged in some spatial pattern so as to enhance the probability of detection of weak seismic signals. In addition, an array provides an initial location of the source of the signals detected. The primary seismic network is supplemented by an auxiliary seismic network of 120 stations, which provide data to the IDC upon request. Data from these auxiliary stations are used to improve on the locations of events determined by the primary seismic network, and also to provide further characterization of these events. Data from auxiliary seismic stations are quite frequently called upon, and about 30% of the signal arrival times used to define events reported in the Reviewed Event Bulletin (see Section 6.2.3) of the IDC are from auxiliary seismic stations. For each event, the IDC primarily requests data from auxiliary seismic stations located in the vicinity of the event.

The IMS component for hydroacoustic monitoring of the oceans is composed of six underwater hydrophone systems, one in each of the world's large ocean basins, and five seismic stations deployed on small islands designed to detect T-phases, which are seismic signals converted from acoustic signals in the

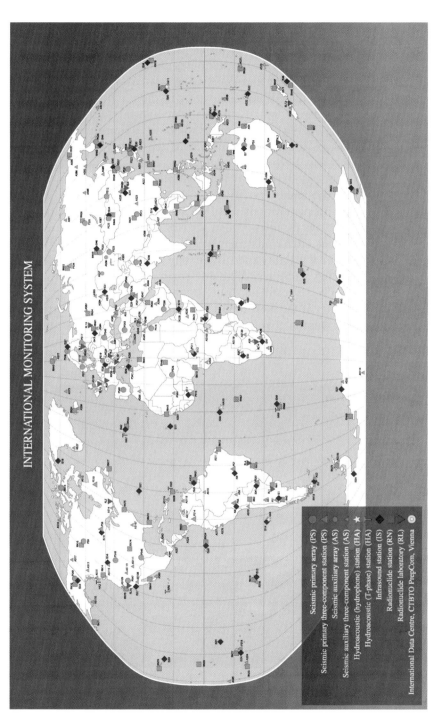

Fig. 6.2 Map showing the distribution around the world of the 321 stations of the International Monitoring System for seismic, hydroacoustic, infrasound and radionuclide monitoring. Also shown are the 16 radionuclide laboratories which support the radionuclide monitoring stations in the analysis of samples of particular interest

ocean at steeply sloping coastlines. Acoustic monitoring of the atmosphere is
accomplished by the 60 infrasound stations of the IMS. These are of the array
type, comprising either four or eight elements each. The establishment of a
global network of infrasound monitoring stations is unprecedented, and there is
considerable scientific excitement and expectation related to the potential of
this network to study sources of acoustic waves in the atmosphere and the
atmosphere itself. A network of 80 stations is monitoring for particulate
airborne radionuclides. Forty of these stations will have equipment to monitor
for radioactive noble gases at entry into force of the Treaty. Radionuclide
monitoring is the only way to establish whether an observed event is nuclear,
and again, the notion of a global network is a novel one, and expectations in the
scientific communities are high.

   Figures 6.3–6.10 show pictures taken at IMS stations, radionuclide laboratories and communications facilities around the world. Stations are located in
all sorts of environments, often climatically hostile, and due care must be taken
to secure their operational stability and sustainability. Stations also need to be
adapted to local building traditions, codes and environmental restrictions. As
these pictures hopefully serve to illustrate, establishment of the IMS is a
formidable challenge, yet at the same time a fascinating experience of cooperation worldwide for a common goal.

Fig. 6.3a Photo of the establishment of the auxiliary seismic station AS73 at Jan Mayen
(Norway), a remote island in the Arctic. The Beerenberg volcano is seen in the
background

Fig. 6.3b During the site survey and installation of the infrasound station IS14 at Juan Fernández Islands (Chile) mules were used to transport equipment and supplies to remote sites. This specific island is know as Robinson Crusoe (since 1966) after the book by Daniel Defoe. Defoe was inspired by the story of the sailor Alexander Selkirk who was abandoned on this island in 1705 for four years and four months

The IMS uses state-of-the-art technology and equipment, whether commercially and generally available, or specially designed for the IMS. The IMS does not in principle deviate from other networks, mostly seismological, established in various parts of the world. The IMS does, however, include some additional

Fig. 6.4a Primary seismic station PS21 located near Tehran in Iran. Such a station could also contribute to the study of large, damaging earthquakes in Iran and its neighboring countries. This station is situated in a hot, dry climate with large temperature differences between winter and summer

Fig. 6.4b Primary seismic station PS42 Kesra in Tunisia. This station is close to the Mediterranean Sea where strong, devastating earthquakes occur

Fig. 6.4c Primary seismic station PS25 at Songino, Mongolia. This is one element of the short-period seismic array. A site survey determined that this location for this new array was better suited than the original Treaty location some 37 km away, from considerations of topology, security and cultural noise. It is also near the existing infrasound station IS34 and the planned radionuclide station RN45, so infrastructure such as power lines and communications can be shared

Fig. 6.5a Auxiliary seismic station AS40 at Lembang, Indonesia. Seismic stations like this one could positively contribute to networks for tsunami warning. The station is situated north of Bandung on West-Java, one of the larger numerous islands in the Indonesian archipelago. The station is situated in a hot tropical climate

Fig. 6.5b Auxiliary seismic station AS70 at Raoul Island, New Zealand. Seismometers are placed in the vault. The satellite antenna provides for transmission of data to the International Data Center in Vienna

Fig. 6.5c Auxiliary seismic station AS106 at Palmer Station in Antarctica. It goes without saying that Antarctica not only has a severe polar climate, but is also very remote

features in support of the political mission of the system, such as the authentication of data to provide assurance to all stakeholders that the data have not been tampered with in any way. IMS stations are also equipped with intrusion detection devices, to monitor physical access to IMS stations, and additional power systems to provide backup if the primary power sources fail. These

Fig. 6.6a Radionuclide station RN33 near Freiburg, Germany

Fig. 6.6b Radionuclide station RN47 near Kaitaia, New Zealand

features contribute to the high quality of the data from the IMS in terms of integrity, reliability and availability, to meet the need for confident verification of compliance with the Treaty. This also makes the data valuable for the purposes of the international scientific community.

Fig. 6.6c Radionuclide station RN49 in Spitsbergen, Norway is housed in this building. The air inlet for the particulate station is seen on the roof of the building

Fig. 6.6d Penguins near the radionuclide station RN73 at Palmer Station, Antarctica

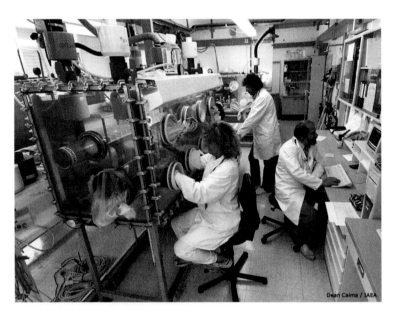

Fig. 6.7 The IMS radionuclide laboratory RL03 is located in the vicinity of the village of Seibersdorf (Lower Austria) about 35 km southeast of Vienna on the premises of the Austrian Research Centers

Fig. 6.8a Communications equipment for the hydroacoustic T-phase seismic station HA02 on Queen Charlotte Islands, Canada. The Queen Charlotte Islands chain on the west coast of Canada faces the deep waters of the Pacific Ocean and is therefore a good location for detecting seismic T-phases

Fig. 6.8b Ship laying cables for the hydroacoustic hydrophone array HA03 at Juan Fernandéz Islands, Chile. This array is situated west of the Chilean coast and is overlooking the southern Pacific Ocean

Fig. 6.8c Landing area of the cables of the hydrophone array HA04 at Crozet Islands, France, one of the most remote sites in the world

Fig. 6.9a Digging cables for the infrasound array IS11 at the Cape Verde Islands. Any array (seismic, infrasound or hydroacoustic) uses many kilometers of cable

Fig. 6.9b Noise reducer for one element of the infrasound array station IS50 at Ascencion, United Kingdom. The inlets are covered with *small piles* of *crushed rock* to reduce the influence of the direct impact of the wind at this windy island site

The Global Communications Infrastructure, connecting IMS stations with the IDC and the IDC with National Authorities, or more commonly their National Data Centers, was put in place from 1998 onwards by a contractor selected on the basis of competitive tendering. This contract will expire in 2008, and a new contract for the following 10 years was already in place in early 2007, allowing for an orderly transition between the two contractors. The communications infrastructure is designed to meet political objectives of transmission

Fig. 6.9c Instrumentation container of the infrasound station IS55 at Windless Bight, Antarctica. As the name suggests this is indeed a low wind site in Antarctica, which is an asset in infrasound measurements. Transportation to the other array elements is done with specialized vehicles

Fig. 6.10 Communication is an integral part of the verification system. At this communications hub the dishes point in all directions

of data and products to States Signatories in a secure, reliable and timely manner. It is largely based on satellite communications technology and the use of dedicated end-to-end connections, but secure internet is also being used where available and adequate. A certain redundancy is built into the system to allow for fast recovery after periods of heavy loading and to allow rerouting of traffic after instances of partial system failure.

The USA established and operated a Prototype International Data Center (PIDC) in Washington D.C. as a contribution to GSE's GSETT-3 experiment, as described in Chapter 3. The IDC in Vienna was initially established in 1997/1998 more or less as a copy of the PIDC. Software was provided from the PIDC, as a gift from the USA to PrepCom, and hardware was acquired via PrepCom's regular budget. The PIDC continued its operation for a few years in parallel with the IDC, and there was extensive cooperation between the two centers, which were linked with a high-capacity line to enable both centers to have access to all available data from new IMS and existing, IMS-designated stations. These arrangements and the close cooperation with the PIDC gave the IDC a flying start. Based on its hardware and software infrastructure, the IDC could also engage in development work from an early date. The IDC could start meaningful on-the-job training activities for its own recruitment purposes, as well as for personnel working at National Data Centers. This personnel will assess the products received from the IDC to see if there are events that need further clarification. Figure 6.11 shows a picture of the IDC environment in the Vienna International Center.

Fig. 6.11 The picture is from the Operational Center of the International Data Center in Vienna. From here, performance of IMS stations is closely monitored in close cooperation with station operators around the world

6.2 Building the Monitoring System – A Gradual but Slow Process

As we have seen, the early enthusiasm led to widespread belief that the monitoring component of the verification regime (IMS, IDC and the Global Communications Infrastructure) could be built within some three years and be ready for its future mission by 2000 or 2001. Current (late 2007) projections indicate that close to 90% of the IMS stations will be in place and contributing data to the IDC by the end of 2008. Why is this taking much longer than originally anticipated? A number of factors and circumstances, political, technical and administrative, have contributed to this, as we shall see in the following.

6.2.1 Political, Cost and Capacity Factors

An assessment of the pace of the build-up of the IMS and the IDC must recognize that the mood and sense of urgency changed with the US Congress decision in October 1999 not to ratify the Treaty. After this decision, a widely shared attitude among States Signatories, clearly repeated by many delegations in every PrepCom plenary debate on the budget for the following year, has been to adapt the pace of establishment of the verification regime to the prospects for entry into force of the Treaty. The date for the completion of the verification regime might thus appear to be a moving target, which has a negative impact on the planning for completion of the verification arrangements. The situation is in fact better than it may appear, though: States Signatories generally subscribe to the view that the implementation of the verification arrangement should increase preparedness for meeting future needs, i.e., the entry into force of the Treaty. States Signatories thus foresee an acceleration of the implementation of the remaining elements of the verification regime when a date for entry into force is established, and States Signatories want to have the best possible point of departure for that final, accelerated phase.

Another limiting factor in the establishment of the monitoring system has been the capacity of the PTS to absorb the work load. The number of staff increased gradually during the first years, reaching a peak of 272 in 2002, after which the level has been relatively stable (see Fig. 6.12 on PTS staff growth over time), and this has enabled the PTS to successfully take on more work. It was evident that the sense of urgency and enthusiasm of the first years also positively influenced the mentality of PTS employees, who were most dedicated to their individual tasks and worked long hours to push towards the establishment of the verification regime. When some key States did not ratify the Treaty and it became apparent that the Treaty would not enter into force any time soon, the enthusiasm among the PTS staff and also among States Signatories' representatives sank to a certain extent. Nevertheless, the ambitions of all stakeholders, as expressed in annual programs and budgets, have been high, and the staff of the PTS has been kept very busy over the years with its tasks.

During the CTBT negotiations, the total investments needed for IMS were estimated at $82 million. As we know, the experts advising the negotiators were overly optimistic with respect to the use of existing facilities in the IMS.

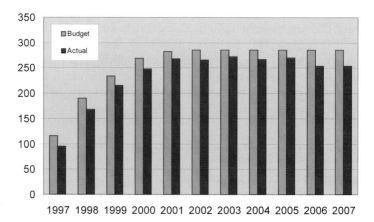

Fig. 6.12 Diagram showing the actual number of staff at the Provisional Technical Secretariat during 1997–2007, along with the number of staff approved in the annual budgets. The latter number has stayed at 286 since 2002

Following PrepCom approval of IMS station requirements and specifications, the PTS estimated in January 1999 that total capital investments of the order of $135 million were required. This figure increased further over the years and stood at $285 million in 2004. The current estimate (late 2007) for total IMS investments is $328 million. Needless to say, these increases have had an impact on the pace of the establishment of the IMS. The annual budgets have, with some very minor variations, been subject to zero real growth since 2003. With increasing costs for IMS station operation and also constant or increasing costs in all other programs, the only remaining option within the constant budget framework has been to slow down the pace of investment in the IMS.

6.2.2 Technical and Administrative Aspects

The global demand for advanced instrumentation of the kind used in the IMS is not very great. Hence, equipment for IMS stations is typically provided by a few, rather small, highly specialized commercial companies with limited production capacity. This often means that delivery times for equipment become fairly long, and the whole process of IMS establishment is, to a certain extent, vulnerable in its dependence on the wellbeing and solidity of these companies. Special IMS requirements, such as authentication of the data, have placed additional demands on the equipment providers and prolonged the procurement processes.

To fulfil requirements and specifications agreed by Working Group B and approved by the PrepCom plenary in 1997, there was a need for substantial equipment development work within some of the IMS technologies. We learned that solid technical work within political environments, with rigorous review and feedback procedures, takes time. So, in spite of the sense of urgency, it was a clear and commonly shared understanding that an orderly process was to be followed, to ensure that all investments in IMS equipment would result in monitoring stations that fulfil their mission. For the infrasound and radionuclide stations, it was necessary to embark on vigorous programs to design and develop equipment and systems for use in the IMS. This was due to a general lack of commercially available, mission capable equipment and also a shortage of internationally shared knowledge and experience. For the radionuclide network, the greatest challenge was in the development of equipment for noble gas detection. An experiment termed the International Noble Gas Experiment was set up and funded from PrepCom budgets to design and develop four different systems; one French, one Russian, one Swedish and one US, which were first tested in parallel at one independent laboratory site in Freiburg, Germany and later deployed at four different IMS radionuclide station locations (Tahiti for the French system, Buenos Aires for the Russian system, Spitsbergen for the Swedish system, and Guangzhou for the US system). For the seismic and hydroacoustic networks of the IMS, the installation of new stations was more straightforward, and was able to profit from commercially available equipment. The establishment of the seismic networks also profited from 20 years of GSE work and a longstanding, strong culture of international cooperation in seismology.

A large number of both primary and auxiliary seismic IMS stations existed prior to the CTBT negotiations. Their designations as IMS stations were based on the recognition of their performance and contribution in a global monitoring context, as evidenced for many of these stations through the experiments conducted by the GSE. But to comply with the IMS station specifications, all these "legacy" stations need to be upgraded technically, with respect to station instrumentation, while some stations also needed attention to the station infrastructure. These upgrading activities soon proved to be very demanding and time-consuming for both the PTS and operators of the existing stations. Even if it is the responsibility of the PTS to upgrade these stations so that they meet IMS operational requirements, the PTS also wants to take into account wishes for the upgrade expressed by station operators. Specific problems relate to the upgrading of the auxiliary seismic stations. Firstly, these stations were promoted during the CTBT negotiations as part of the IMS on the premise that many of them already existed and that the need for capital investment was minimal. The PrepCom, however, decided that the specifications for these stations should be the same as for the primary seismic stations, resulting in a substantial need for upgrading. The pre-existing auxiliary seismic stations are often stations of some regional or global scientific network and need to continue operating in that capacity, even after the introduction of these stations into the IMS. This led to sometimes complicated discussions with network operators to ensure that such stations would fulfil their function both as IMS stations and also as elements of the scientific network. These stations would thus respond to requests by the IDC for data without delay and without any conflict arising from their dual use. In some instances this could be resolved with the network operator by introducing a station interface that could accommodate the dual use of data; in other instances it was necessary to install a complete, new set of equipment next to the pre-existing station in order to avoid problems related to dual use. All this means that the installation of auxiliary seismic stations has taken more time and resources than anticipated.

The PTS legal and procurement functions have been heavily involved in the processes of establishing the IMS by concluding agreements with host States on site surveys and station building, and with contractors for delivery of equipment, hardware and software, establishment of infrastructure, and station operation and maintenance. Limited capacity in these functions at times delayed the signing of contracts with providers of equipment and services, as did the rather rigorous rules and regulations adopted for the procurement process. Complicated PTS procedures as well as different administrative and legal frameworks in hosting States, as we shall see below, have been serious hurdles in the establishment of the IMS.

As we have seen in Chapter 5, the signature of facility agreements with the PrepCom or the exchange of letters to allow station installation to proceed, has been a slow process in many States Signatories. Difficulties have ranged from legal, taxation and customs issues, to the identification and designation of relevant technical contacts in the various countries. Even for countries in which all formal arrangements with the PrepCom are in place, there may be problems finding solutions to IMS station installation issues with local authorities, such as issues related to access to and use of land. For example, it has been difficult to secure land for a planned IMS infrasound station in northern Norway. Such stations require that some pieces of land are fenced to protect steel pipes that are placed on the ground for the purpose of noise reduction, and to avoid these pipes causing harm to people and animals that inadvertently might move across them. In the Norway case, such an installation, with its fences, may represent an obstacle to the traditional seasonal migration of herds of reindeer. This factor was taken into account by the local council in its decision not to permit establishment at the location that was deemed to be the most suitable one for this station. So preparedness, but also local conditions and circumstances in 89 IMS hosting States are crucial to the IMS installation process.

6.2.3 IMS and IDC Status Overview

In spite of all of the factors described above affecting and slowing down the work on the IMS, its establishment has been on a fairly steady path over the years. Figure 6.13 shows the development over time in terms of completion of

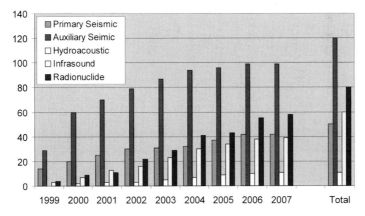

Fig. 6.13 Diagram showing the progress during 1999–2007 in the installation of IMS stations of the various technologies. As can be seen, the station establishment has been on a very steady course over the years towards completion of the IMS. Total refers to the number of stations defined in the Protocol to the Treaty

installation of stations for the various technologies. As we have seen, it is expected that by the end of 2008, close to 90% of the stations of the entire IMS will be in operation and sending data to the IDC. The installation of the remaining 10% will require host country specific solutions, and several of these stations are in States that have not yet signed the Treaty. The Treaty states that *"At entry into force of this Treaty, the verification regime shall be capable of meeting the verification requirements of this Treaty"*. The text on the establishment of the PrepCom states that the PrepCom *"shall undertake all necessary preparations to ensure the operationalization of the Treaty's verification regime at entry into force, pursuant to article IV, paragraph 1, and shall develop appropriate procedures for its operation, presenting a report on the operational readiness of the Treaty's verification regime, together with any relevant recommendations, to the initial session of the Conference of the States Parties"*. Even though the PrepCom may have done its utmost, all 321 IMS stations may not be in place at the time of this initial session of the Conference of the States Parties. Nevertheless, the Treaty will be in force at this time, and this conference may then decide to accelerate efforts to complete the IMS.

As we have seen, the IDC got off to an early start in Vienna. A commissioning plan for its development through seven phases was adopted at an early date and called for a mission-ready IDC for operation following entry into force within a matter of some three years. However, political signals soon prompted an adjustment to new goals and timelines, also for work at the IDC. The IDC staff started on an orderly process to develop ownership of the software, add functionality and routines that were not included in the gift from the PIDC, and migrate applications software towards open source solutions, including the Linux operating system. It has further conducted studies, often with the help of contractors, to enhance the performance of the automatic and interactive data processing tools to detect and locate events.

In parallel with its development activities, the IDC has focused on providing States Signatories with products from its operations. From 20 February 2000, the IDC has produced a daily Reviewed Event Bulletin, largely based on seismic data but also with some contributions from signals detected on the hydroacoustic and infrasound stations. The IDC is also producing a daily Reviewed Radionuclide Report. In these products, IDC analysts use their expertise to interactively review and improve upon products from the automatic processing cycles. The IDC has conducted a considerable amount of training activity, with a view to recruitment of its own personnel, but also to the benefit of personnel at National Data Centers around the world. In support of all of these activities, the IDC has established a computer infrastructure that is operating almost without outages for all internal and external services. As of 2007, the Reviewed Event Bulletin contains on average 80 events per day. For radionuclide particulate stations, about 40 spectra are automatically analysed, interactively reviewed and categorized every day, with results provided in the Reviewed Radionuclide Reports. These reports are unique in the sense that they represent the only effort to issue analysis results regularly and with short delay, based on a global network of radionuclide stations. Software for analyzing noble gas data has been developed and is now in use for the two principal types of noble gas detection systems installed at IMS radionuclide stations.

The political situation has thus not only prevented the Treaty from entering into force, it has also slowed down the implementation of the verification regime. It would have been possible to complete the verification system in a shorter time, given the political will, sufficient resources and full cooperation by all countries hosting IMS stations. The slow pace has, however, provided an opportunity to use ample time to develop concepts and ideas, find common ground among States Signatories on key issues such as use of the noble gas technology, and concentrate on training activities worldwide. The slow pace also allows more thorough testing and evaluation of the systems. There is also a need for States to further develop their capability, nationally or in regional cooperation, to interpret and assess in Treaty monitoring terms the information provided by the Treaty verification arrangements, as we shall discuss in Chapter 8.

6.3   Some Specific IMS Issues

In several areas, the Treaty goes into quite some level of detail in its description of verification-related arrangements, but it still leaves a range of technical issues to be worked out and resolved during the PrepCom phase. The Treaty often refers to requirements that are to be specified in relevant operational manuals. To fill the manuals for the various IMS station technologies with the required detail and otherwise make progress towards the completion of the IMS, a series of issues have been and still are being dealt with in the PrepCom and its subsidiary bodies, in particular Working Group B. In the following we dwell on some issues that have been important from a policy point of view. In some cases the resolution of these issues presented challenges to the consensus forming process.

6.3.1   Station Specifications and Certification Procedures

As we have seen, the PrepCom had already agreed on technical station specifications in 1997, to allow IMS station establishment to proceed without delay, and these technical specifications have remained unchanged to this date. The next related issue was to define requirements that must be fulfilled by an IMS station during the PrepCom phase, to provide assurance that it would fulfil its mission after entry into force of the Treaty. These requirements are formulated in procedures for IMS station certification approved by the PrepCom. For a station to be certified, its site, the station equipment and its infrastructure must substantially meet the agreed technical specifications for IMS stations. In addition, data authentication devices must be in place and must have been demonstrated to function properly, and the station must be connected to the IDC via a communications link that has also been demonstrated to work properly. Once a station certification is granted, the station operator is entitled to have station operation funded via the PrepCom budget (except for auxiliary seismic stations, see Section 6.3.2). This linkage between certification procedures and future funding of station operation prolonged the discussions in Working Group B over the certification requirements. However, agreement was reached in May 1999, as delegations became comfortable that certification requirements were set at such a high level that it was justified to spend PrepCom's money on the operation of certified stations. The first certifications to IMS standards were made in July 2000.

The approved specifications for IMS stations are met by equipment provided by a variety of manufacturers, and over the years products from a range of such manufacturers have been installed at IMS stations. The PTS has attempted to standardize some of the equipment through their procurement policy, using so-called call-off contracts containing options for future purchases of additional units. Upgrading stations typically involves retaining many of the existing components of the stations to save money, at least in the short term. Then there are IMS stations for which the host State selects the equipment. These are the stations built under the arrangements referred to as "reduced assessment", whereby a host country meets the costs for station establishment, and is compensated by a corresponding reduction in its assessed financial contribution to the PrepCom. In such cases it will be the responsibility of the host State to make sure that the station is certifiable. This general lack of standardization of station equipment will not prevent the IMS from fulfilling its mission, but may present challenges to effective and economic logistics support and maintenance of the IMS. On the other hand, keeping several suppliers engaged in delivery of equipment to the IMS reduces the dependency on single providers.

Ten years have passed since the first stations were installed by the PTS and there will soon be a need to start replacing equipment and technically upgrading these early stations. This technical re-capitalization of the network will be an extensive and essential undertaking and it will provide an opportunity to further standardize the equipment to improve the cost-effectiveness of the operation.

6.3.2 Auxiliary Seismic Stations

Auxiliary seismic stations of the IMS transmit data to the IDC only upon request, as opposed to primary seismic stations, which transmit uninterrupted data on-line to the IDC. The main rationale underlying this is that the auxiliary seismic stations are to provide data that will reduce the uncertainty of the location of events already detected and initially located by the primary seismic stations. Mechanisms and protocols have been put in place and are being used by the IDC to request selected segments of data from auxiliary stations for use during automatic as well as interactive stages of the bulletin production. At times, data from newly installed auxiliary seismic stations have been transferred in continuous mode to the IDC for technical testing and evaluation. It has been argued that there would be benefits from making such arrangements permanent and that the capacity of the communications infrastructure is or can be made sufficient to accommodate the additional data. It has even been suggested that it might be more convenient and perhaps even less costly to routinely transfer data from all auxiliary stations to the IDC and then to select what is needed in the analysis. This situation is in contrast to the realities at the time of the CTBT negotiations, when there was good reason to limit the volume of data to be transmitted, purely for cost implications. There might thus be neither technical nor cost reasons why auxiliary data could not be more fully used in the analysis. The reason is purely political and the current understanding in the PrepCom is that the use of auxiliary seismic data will be as initially intended. It might, however, be interesting to study how a more extensive use of auxiliary data might facilitate the analysis and improve on the quality of the products. For example, letting the auxiliary seismic stations take part in the event detection stage of the data processing would lower the event detection threshold of the seismic component of the IMS. Any issue on the use of auxiliary seismic stations after entry into force of the Treaty will, however, need to be resolved by the initial session of the Conference of the States Parties when that body approves the respective operational manuals, or by a later Conference of the States Parties.

The IMS auxiliary seismic stations have a special status in the Treaty compared to the stations of the primary seismic, hydroacoustic, infrasound and radionuclide networks: the Treaty text implies that operation and maintenance of these stations will not be covered by the CTBTO. Hence no provision has been made to accommodate such costs in the annual PrepCom budgets. Most host countries have been prepared for this and have made appropriations in their national budgets for such costs, and some stations are provided for through their dual use or association with some other regional or global network of stations. For a few auxiliary seismic stations, though, operations have come to a halt due to lack of funding. These are new stations, established in developing countries at locations where no appropriate facilities were in existence prior to the Treaty negotiations. So far no PrepCom money has been provided for operating auxiliary stations as the PrepCom, for policy reasons, does not want to make exceptions to the funding arrangements. The PTS is now asking host countries for their commitment to fund operation and maintenance of auxiliary seismic stations on their territory before investments are made. The basic funding problem remains unresolved, however, as far as some host States are concerned, and it may become necessary to revisit the issue, with the aim of securing the proper functioning of all stations of the auxiliary seismic network.

installation program to proceed. These procedures have served the PrepCom well, and nearly every session of Working Group B since 1998 has more or less routinely dealt with coordinate changes proposed by the PTS. A coordinate change is dealt with in the Policy Making Organs if the coordinate change involves a distance exceeding 10 km. For changes exceeding 100 km, the PTS is required to present technical evidence in the form of capability modelling that the overall performance of the monitoring network is not adversely affected. There have been very few controversies in the PrepCom over such moves, even when stations have been moved hundreds of kilometers. There is one unresolved issue, though, which relates to a proposed move of a station from one State Signatory to another.

6.4 On-Site Inspections a Politicized Issue on a Slow Path

To deal with concerns about possible non-compliance after entry into force of the Treaty, any State Party has the right to engage other states and the CTBTO in a consultation and clarification process and eventually request an on-site inspection (OSI), as explained in Chapter 4. Arrangements for these elements of the verification regime need to be developed, established and tested during the PrepCom phase, so that they are in place and will function properly after the Treaty has entered into force. To prepare for the consultation and clarification element is mostly a matter of developing and documenting procedures. The development of the OSI regime involves, among other things, elaboration of procedures for the conduct of an OSI in the form of an OSI Operational Manual, development and testing of a training program for future inspectors, and specification, development, acquisition and testing of equipment for training purposes. Figures 6.14a and b show pictures of field activities conducted for the development of the OSI regime.

It was an achievement in itself to reach agreement during the CTBT negotiations on a fairly comprehensive and elaborate regime for on-site inspections, considering the intrusive nature of the arrangements involved. Only few participants in the negotiations had prior knowledge and experience regarding this kind of on-site inspection. Inevitably, compromises were struck with the understanding that further elaboration would be needed to fill the OSI regime with the required details for its proper functioning. So, in spite of the long and elaborate text on the OSI in the Treaty and its Protocol compared to other elements of the verification regime, much work to fill in additional details was left for the PrepCom phase.

A real on-site inspection can take place only after entry into force of the Treaty. This is different from the circumstances under which the IMS/IDC monitoring system is being developed. PTS here has the responsibility to develop and establish a real system that is put to realistic tests every day by naturally occurring events. This brings a clear focus to the work. Such a clear

Fig. 6.14a Participants boarding an M18 helicopter for visual observation and gamma survey during an on-site inspection exercise in Kazakhstan in 2005

focus was for a long time lacking in the OSI work, both in Working Group B and the PTS.

The PTS started its OSI work with a comparatively small number of staff, which grew to about 20 by 2002 and has since then stayed at this level or a little

Fig. 6.14b During this OSI field exercise soil samples are taken for further analysis in a laboratory

above. It soon proved difficult to get started in a well-organized way, both in the PTS and the Policy Making Organs. This was due partly to the absence of an overall plan for the development of the OSI, but also to the political sensitivities associated with this component of the verification regime. States participating in the OSI work in Working Group B had very different perspectives on the work to be done, and differing national priorities. For political reasons, States Signatories advocated that work on a range of OSI issues, including highly technical ones, should be conducted in Working Group B by their national delegates, rather than leaving issues to the PTS. As we saw it, many States deeply engaged in the development of the OSI regime had a tendency to find sensitivities in most issues dealt with in that context. Sensitivities over issues such as technical specifications for fairly simple equipment for use during OSIs sometimes resulted in protracted discussions in Working Group B before agreement was finally reached. Again, the underlying factor and concern is the intrusiveness of the OSI regime and its potential for uncovering sensitive information outside the scope of the Treaty, although the Treaty emphasizes the right of an inspected party to protect sensitive installations not related to the purpose of the inspection. The elaborations were further complicated by the fact that the Working Group B participants had different levels of experience of comparable activities, and it is apparent today that the scope of technical issues to be tackled were not very well understood initially. It was also realized after some time that "off the shelf" equipment and methods could often not be applied to the OSI work. Examples here are the requirement for "blinding" of radionuclide detectors and the non-existence (until recently) of mobile field equipment for detecting radioactive noble gases.

The Treaty states that the initial session of the Conference of the States Parties shall consider and approve a list of equipment for use during OSIs. To develop this list, equipment has been acquired, tested and used for training and other purposes under field conditions. Other equipment belonging to institutions in States Signatories has been loaned to the PTS for various experiments. States have also committed some equipment for future use by the PTS, to relieve the PrepCom of the expense of investing in equipment that would incur maintenance costs and later be technically outdated and in need of replacement. In addition, PTS has worked on the development of methodology and procedures for the conduct of an OSI.

In August 2001, the USA, following review by its new administration of the country's position on nuclear issues, withdrew from all PrepCom activities related to the OSI regime. Until then, the USA along with the Russian Federation had been the driving force for the development of the OSI regime, based on the extensive, relevant experience and knowledge in these two countries. To lose this major player and its expertise in discussions in Working Group B was a setback to the process of developing the OSI arrangements. This also raises the question of which revisions will be needed when the USA at some future time re-enters the scene. This question may also be asked in more general terms in relation to other currently non-signatory States that might enter the PrepCom process at some time in the future.

The challenge for those working in the OSI field is to prepare in the best possible way for circumstances that during the PrepCom phase can at best be simulated to credibly resemble some future scenario. Under these circumstances the PTS and the Policy Making Organs made slow progress on the OSI regime for several years. A number of activities were pursued, however. Workshops and general and specialized courses for representatives of States Signatories have contributed to the training program for future inspectors, and approximately 500 individuals have participated in OSI activities and are in a roster of experts to participate in future advanced OSI activities. Not all of these individuals may be available for such activities, though, as there is no formal commitment to remain available during the PrepCom phase.

In an external evaluation of the OSI activities (CTBT/WGB-21/INF.5 2003) conducted in 2003, that is, after six years of preparatory work, the international experts team *"acknowledges that planning is undertaken within the OSI Division in various forms but considers that this falls short of the overall planning required to achieve an operational OSI capability"*. The team continues *"Of particular significance, the Team considers that planning and development should be subject to the discipline of a specific goal, that of a fully comprehensive field exercise, trialling OSI at near full scale, to be implemented no later than 2007"*. There was obviously a need to focus activities on a challenging exercise. In 2005 PrepCom decided that a large scale field test, including many elements of a real on-site inspection, should be conducted in 2008. This test is discussed further in the next chapter. The planning and preparation for this exercise provided the necessary impetus and direction, and the PTS is now well underway to developing a clear understanding of what it takes to conduct an on-site inspection.

A major element of the OSI regime is the Operational Manual. This is the document that will detail procedures for the implementation of future OSIs and guide their conduct. It needs to be worked out in draft form during the PrepCom phase and must be approved by the initial session of the Conference of the States Parties. Nearly every aspect of OSI related work in the PTS and the Policy Making Organs finds its way into this manual. There were some initial discussions on how to go about developing it. Given the political sensitivities, States Signatories chose to institute a process under the auspices of Working Group B. The role of the PTS has been to support this process with draft text for technically oriented and less politically sensitive sections of the manual, whenever so requested by the States.

It proved difficult to get this manual development process on the right track and sufficiently focused in Working Group B. It was also difficult to coordinate and integrate the process of drafting the manual with other activities that could provide input to it. After a while, though, a fairly orderly and sustainable process was created, under the guidance of capable task leaders and their "friends". The work on the OSI Operational Manual represents a huge collective effort on the part of all States that have participated in this work, as well as the PTS. Five to six weeks have typically been spent in meetings of the task group during sessions of Working Group B in Vienna each year, and in addition, drafting activities, including special workshops, have taken place intersessionally. Methodology and procedures have also been tested in various forms of experiments, ranging from table-top to large-scale field exercises. Over the years, text covering all aspects of an OSI has been created and elaborated by the task group for this manual, and the process is currently (late 2007) nearing completion of the second reading of the text. The draft OSI Operational Manual currently comprises more than 1000 pages of text, which still contains a large number of brackets reflecting that not all is yet agreed. The development of the Operational Manual also received a much needed boost with the decision to conduct a large scale exercise in 2008. A most significant step forward was taken by agreeing on a fairly comprehensive Test Manual to be used for that exercise. This is addressed further in the next chapter.

Chapter 7
Testing Shows High Performance

7.1 From Building to Testing

In the early PrepCom years, IMS station building dominated the activities, both in the PTS and in Working Group B. The Working Group delegates' keen interest in this activity is readily understandable, as station establishment is easily grasped by everyone. Moreover, hosting States generally derive national pride from having IMS stations on their territory. Station establishment involves work in 89 different countries, and it also demonstrates measurable progress toward completion of the verification system. In fact, the number of certified IMS stations has, for better or worse, become the most tangible metric for the assessment of progress in this regard. In parallel with the station building activities, PTS staff worked steadily on the establishment of the IDC and training of its future analysts, although these activities did not catch the attention of the States Signatories in the same way. Gradually, however, the emphasis shifted from these build-up activities to the operation of the stations and the IDC. The Treaty states that all IMS stations shall be owned and operated by the States hosting them. Operation at the IMS station level is typically handled by a station operator under a contract with the PTS and thus involves many institutions and people in States Signatories around the world. A growing number of people at the PTS also became involved in operational tasks over the years, such as checking the quality of incoming IMS data, running automatic data processing "pipelines", interactively reviewing the results of the automatic data processing, and disseminating data and products to States Signatories.

7.1.1 Mode of Operation of the Global Monitoring System

Initially, the ambitions for IMS station operations and IDC processing were very high. The requirements for station operation, as reflected in contracts between the PTS and station operators, were the standards agreed for operations after entry into force (EIF) of the Treaty. Reviewed Event Bulletins were

O. Dahlman et al., *Nuclear Test Ban*, DOI 10.1007/978-1-4020-6885-0_7,
© Springer Science+Business Media B.V. 2009

to be completed within about two days, in accordance with the text elaborated for the operational manual for the IDC. These lofty ambitions were fairly natural during the first few years, given the expectations of an early EIF, and at least nominally set the norm until late 2001, when a debate emerged in Working Group B over the general mode of operation of the monitoring component of the verification system, including the IMS, the IDC and the Global Communications Infrastructure (GCI), during the PrepCom phase. An understanding was reached in 2002 that *"pre-EIF activities, especially the technical testing and provisional operation and maintenance of certified IMS stations, the GCI and the IDC, do not provide for verification of compliance with the CTBT, including monitoring for verification purposes"*. This mode of operation was already stated in the "Text on the Establishment of a Preparatory Commission", adopted by the States Signatories in 1996. The 2002 decision served to reinforce this original understanding, and also served to point out what is *not* the purpose of activities during the PrepCom phase. At the same time, some of the operational requirements were relaxed. For example, formal contract requirements for data availability for IMS stations approved for post-EIF operations were waived. The IDC staff would have a five day work week with normal business hours, and a relaxed schedule to produce the Reviewed Event Bulletin within 10 days was established.

This important common understanding of the mode of operation during the PrepCom phase, developed in 2002 and restated on an annual basis since then, has brought constructive clarity and has greatly helped to facilitate the work in the PTS and in PrepCom's Policy Making Organs. It is thus accepted by all stakeholders that it is not a requirement that the IMS/IDC/GCI system should be operated at post-EIF standards during the PrepCom phase. It is rather the purpose of the current testing and provisional operations to ensure that the system with all its components develops stage by stage, to meet the post-EIF operational specifications in a cost effective manner at EIF. Other stated objectives of the PrepCom operational activities are that investments made by the PrepCom in equipment as well as human skills should be protected and preserved. The operations should also support the ability of States Signatories to develop their National Data Centers. The challenge is thus to develop and test systems, and evolve towards EIF readiness, without having in place an operational or de facto monitoring system already during the PrepCom phase. Experience so far indicates that the fundamental PrepCom objective of delivering an operational system at EIF can likely be met by a combination of time-limited, system-wide performance tests and well-targeted tests of subcomponents of the integrated system.

In keeping with this mode of operation during the PrepCom period, the daily work must emphasize the lessons learned and feedback to systems development. In this way, and perhaps only in this way, can the system evolve progressively and be ready to fulfil its mission upon EIF. In practice, however, it proves to be a challenge to set aside sufficient resources for feedback, given the daily workload needed to keep all systems operational and the expectations of continuity in delivery of data, products and services. Be that as it may, it is crucial for any system development project to set aside enough time and resources for feedback. It is essential to look back at times, evaluate performance and learn from experience, and then improve the systems. This is essential when dealing with technical components of the system and it is even more important when it comes to the training of IDC analysts and development of IDC analysis procedures.

Still, under the relaxed mode of operation since 2002 and even following the financial strain in 2007 (see Chapter 10), the Policy Making Organs wanted the PTS to produce an analyst-reviewed bulletin of high quality for seismo-acoustic events, based on analysis of seismic, hydroacoustic and infrasound data, for every data day, and also to analyze spectra from radionuclide stations on a regular basis. This is a substantial undertaking in itself, with a workload that increases with the ever-increasing number of IMS stations being used in the analysis. Since 2002, the IDC has issued its bulletin largely according to the agreed schedule, with the exception of some days with a large number of aftershocks following the Sumatra earthquake on 26 December 2004 that generated the disastrous tsunami in the Indian Ocean. This sustained operation has merit in itself, also in the context of system development, given that the experience gained is analyzed, documented and fed back into the development process. A routine operation also demonstrates what resources it takes to produce the daily bulletin, based on the current automatic and interactive tools. Figure 2.5 shows the 27 574 events in the Reviewed Event Bulletin for 2006. This figure shows that the analysis of data from an evolving IMS network very adequately represents the global seismicity, as known from decades of operation of thousands of seismic stations around the world.

With its resources, the PTS is in a position to carry out so-called "special event analysis". Such analysis is mandated in the Treaty, and is meant to assist States Parties, after entry into force of the Treaty, in their consideration of events of possible concern, before States Parties themselves pass their judgement on the nature of such events. Also during the PrepCom phase, States have sometimes requested the PTS to carry out special analysis of specific events. It can be argued that such analysis work by the PTS would, effectively, be monitoring for verification purposes, which is not consistent with the approved, current mode of operation. On the other hand, it is understandable that the PTS would like to show that investments made so far in the IMS and IDC would prove useful to States Signatories in their assessment of events that occur during the PrepCom phase. To clarify this situation and guide the PTS in the conduct of its work, the Policy Making Organs have decided that during the PrepCom phase, such special studies should be carried out for IDC development purposes only, and that the outcome of all such studies will be available to all States Signatories.

7.2 Global Tests Show High Performance of IMS/IDC

The continuous operation of the IMS network, the communications infrastructure and the IDC provides opportunities to evaluate the technical functioning of the system, to assess current global monitoring performance, and to make projections of the eventual capabilities after completion of the IMS.

7.2.1 Technology Works

IMS and IDC operations in the current, provisional mode compare favorably with the performance of other networks in terms of metrics such as station and communication link uptimes, and timeliness of the production of automatic and reviewed bulletins. As part of the ongoing operations, some specific tests have been undertaken, most notably the so-called System-Wide Performance Test 1 (SPT1), performed in several phases during 2004 and 2005. During this test, a wealth of information pertaining to the performance of all systems and sub-systems, hardware and software alike, was collected and analyzed. Statistics on station uptime showed that 60–70% of the 163 stations for the various technologies participating in the 2005 phase of the test met the requirements specified for operation *after* EIF. These requirements stipulate an uptime of 98% for primary seismic, hydroacoustic and infrasound stations, and 95% for radionuclide stations. These results are encouraging, taking into account the relaxed mode of operation agreed for the PrepCom phase, during which status checking and repair services are performed by PTS personnel and station operators during regular working hours only. These figures also show that investment in high quality station infrastructure and equipment is paying off. SPT1 also examined responses to failures imposed under controlled conditions, like a complete shutdown of the global communications infrastructure. All in all, SPT1 was a very useful performance evaluation activity, which revealed high operational performance for many stations but also weaknesses in various areas that require improvement to current practises. SPT1 was also a clear demonstration of the role of testing, from an overall system perspective, in the further development of the monitoring system. Drawing upon SPT1, narrower focused exercises to test individual components of the system have been conducted in 2007 and additional tests are planned for 2008. These exercises might lead to another system-wide performance test at a later time.

7.2.2 Monitoring Capabilities Foreseen During the CTBT Negotiations

When it comes to IMS monitoring capabilities, some crucial questions are: How small an event will the IMS be able to detect? And with what accuracy can the geographical locations of events detected by the IMS be determined? Before attempting to address these questions, we will try to assess what the expectations were at the time the configuration of the IMS in terms of number and location of stations was negotiated in the Conference of Disarmament in Geneva. With a CTBT banning *all* nuclear explosions, it would not have made sense for the CTBT negotiating bodies to provide functional specifications for the detection or event location capability of the IMS, and the negotiators did not do so. The IMS was thus not designed to meet specific requirements in terms of capabilities to detect and locate events. On the other hand, using modelling tools available at the time of the negotiations, any proposed network could be characterized by threshold values, which indicated, for any location on earth, the smallest size of an event that would be detected with a given likelihood by that network. This implies that it was also possible to estimate what size events would go undetected by the network. It was then be up to the individual States taking part in the CTBT negotiations on technical verification measures to pass judgment, based on political, national security and other considerations, on what would constitute, for them, sufficient capabilities for the monitoring network.

Although no decision was made by the Conference on Disarmament on specifications for IMS monitoring capabilities, much of the work of the expert groups formed as part of the CTBT negotiations revolved around considerations of various network options in terms of their capability to detect, locate and identify nuclear explosions of a yield of 1 kt TNT or more in the atmosphere, underwater and underground (CD/NTB/WP.224 1995, CD/NTB/WP.269 1995, and CD/NTB/WP.283 1995). This approach was guided by views expressed by delegations in the early stages of the CTBT negotiations, regarding the adequacy of various options of monitoring capabilities, and provided a common baseline for the evaluation of proposed network configurations for the different technologies. The method of work adopted during the technical discussions was then to propose, analyze and adjust configurations for the networks of the four IMS technologies until consensus was reached on numbers and distribution of monitoring stations in each of the four networks. In this process, different modelling tools were used to calculate expected capabilities of various designs. The result of the use of one such tool is illustrated in Fig. 7.1, which shows, for the agreed IMS primary seismic network, the projected detection capability as calculated by researchers at the Center for Monitoring Research and published by the National Academy of Sciences (NAS 2002). For each location on the entire globe, the figure shows the magnitude of the smallest seismic event at those locations, which would be detected with a 90% probability by three or more stations of the primary seismic IMS network. For events at this magnitude threshold and higher, detection at three stations or more ensures that useful estimates of their geographical locations can be made. Network capability projections like those in Fig. 7.1 are based on assumptions about the distribution in the network of the two station types (array with a capability to suppress noise, or three-component station), background noise

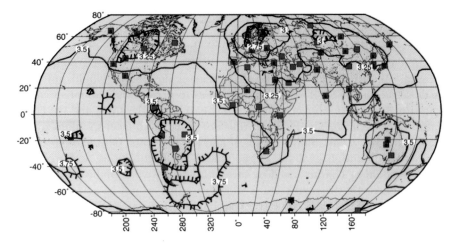

Fig. 7.1 The projected detection capability of the IMS primary seismic network (NAS 2002), as estimated by researchers at the Center for Monitoring Research. For each location on the entire globe, the figure shows the magnitude of the smallest seismic event at those locations that would be detected with a 90% probability by three or more stations of this network. Reproduced with the permission from the National Academies Press, Copyright 2002, National Academy of Sciences

level at each of the station sites, and seismic wave propagation characteristics. For the seismic networks, much of the information needed for capability calculations was already available at the time of the CTBT negotiations, e.g., from experience gained from the GSETT-3 experiment as well as previous experiments conducted by the GSE.

To relate detection capability maps like the one shown in Fig. 7.1 to the yield of underground nuclear explosions expressed in tons of TNT equivalent, magnitude-yield relationships are needed. Such relationships vary with source region, depth of burial of the explosion, and properties of wave propagation paths. Considering nuclear explosions in so-called "hard" rock conducted without any attempts at evasion of detection, various such relations indicate that a 1 kt device will give a seismic magnitude in the range 4.25–4.45 (NAS 2002). Using these relations to infer yields for magnitudes as shown in Fig. 7.1, it is found that for practically all of Europe, Africa, Asia and North America the detection threshold of underground nuclear explosions is below 100 tons of TNT in hard rock. It should be emphasized that the magnitude-yield relationship would be different for explosions that are not conducted in hard rock. Explosions in water or clay couple very efficiently into seismic signals, so the explosive yield producing a certain seismic magnitude would be even smaller than for the hard rock scenario. Conversely, explosions in soft rock, like unconsolidated dry sediments, couple less efficiently into seismic signals. Other magnitude-yield relationships would also apply to explosions conducted evasively, for example in underground cavities.

The papers of the expert groups of the CTBT negotiations are not explicit when it comes to the geographical variations in the detection capability of the primary seismic IMS network. It was stated in CD/NTB/WP.224 (1995) that this network would have *"an ability to detect and locate explosions at the magnitude 3 ½ level, equivalent to an explosion of the order of 0.5 to 1 kiloton"*. Strictly speaking, this statement applied to a proposed 46-station network, which was later supplemented by four additional stations, resulting in *"slightly better capabilities"* (CD/NTB/WP.269 1995) than those projected for the 46-station network. The expert group papers thus provide one overall number for the detection capability, presumably representing a global average, which would be fairly consistent with the findings shown in Fig. 7.1 from the study for the same network issued by the National Academy of Sciences (NAS 2002). The expected capability in terms of explosive yield differs, however. According to the National Academy of Sciences (NAS 2002), a seismic magnitude of 3.5 is believed to correspond to a yield of only 100 tons (0.1 kt) of TNT in hard rock. The CTBT negotiators were thus conservative in their statement quoted above, in the sense that they also took account of explosions in other environments than hard rock, with less efficient coupling into seismic signals.

Regarding the location capability of the seismic component of the IMS, however, the expert group papers from the negotiations are a little more explicit. As we have seen, the main purpose of the IMS auxiliary seismic network of 120 stations is to provide data to improve on the accuracy of the event locations derived by the 50-station primary seismic network. In CD/NTB/WP.283 (1995) it is stated that *"The resulting combined primary-auxiliary network should, in the view of the Group, be capable of locating seismic events in continental areas and coastal regions of magnitude 4 and higher with an uncertainty of less than 1000 square kilometers,..."*. This statement applied to a proposed 119-station network, which was later modified slightly before final adoption of the 120-station IMS auxiliary seismic network, but the projected location capabilities of these variants of the network are nearly identical. The experts realized that in order to reach this location capability, calibration of the combined primary/auxiliary seismic network would be needed. Such calibration involves mapping of travel times from source to receiver for well-constrained events, like explosions with known origin times and locations, and earthquakes recorded by a network of stations surrounding the earthquakes at short distances. This enables regional corrections to be made to travel times for standard earth models, and will result in significant improvements in the accuracy of event locations, by reducing the bias. Although not a formal requirement, it is not by chance that the goal of locating events with an uncertainty of less than 1000 km^2, at a 90% confidence level, was pursued in the design of the seismic component of IMS. As we have seen, the protocol to the Treaty specifies that the area of an on-site inspection shall not exceed 1000 km^2.

What capabilities did the experts participating in the CTBT negotiations foresee for the hydroacoustic, infrasound and radionuclide components of the

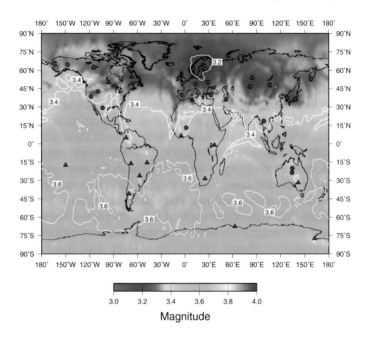

Magnitude

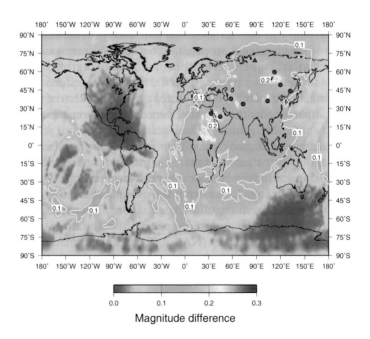

Magnitude difference

An assessment of event location accuracies for the seismic component of the IMS can be guided by operational statistics available for 2006, which show that the locations of only 21% of the 27 574 events in the Reviewed Event Bulletin for this year were determined with an associated uncertainty area of less than 1000 km^2. This shows that there is still quite some way to go before the anticipated performance level is reached. As we saw in Chapter 2, uncertainties in event locations originate from both statistical uncertainties and bias. The uncertainty areas of the event locations published in the Reviewed Event Bulletin account for both of these effects. Further calibration efforts will improve on the bias, and other ongoing work to improve software for the computation of seismic signal attributes used in the event location process, such as the estimate of the direction of arrival of detected signals, is also expected to reduce the uncertainty. Of the events in the Reviewed Event Bulletin, 60–70% have magnitudes less than 4. An uncertainty area of 1000 km^2 or less was anticipated during the CTBT negotiations, as discussed above, for events of magnitude 4 and above, for which the uncertainty area will be smaller than for events below magnitude 4. When the completed seismic network is in place, there will in general be a larger number of station detections for a given event magnitude, which also reduces the location uncertainty. Finally, it should be born in mind that the above uncertainties refer to the current *regular* production mode for the bulletin. Special analysis for challenging events, using all available data and analysis tools, has shown a potential to substantially reduce the event location uncertainty areas.

As we described in Chapter 4, Annex 2 of the Protocol to the Treaty lists a number of characterization parameters to be used by the IDC for event screening. This list of parameters is tentative. All seismic events included in the Reviewed Event Bulletins with a magnitude greater than or equal to 3.5 are subjected to the event screening process. During 2006, 61% of these events were screened out, implying that they are earthquake-like, 28% were inconclusive due to lack of information required to undergo screening, whereas the remaining 11% were actively *not* screened out. These statistics indicate that

Fig. 7.2 The *upper part* of the figure shows the detection capability of the IMS primary seismic network in its current (late 2007) state of development, with 38 stations sending data to the IDC. The capability is represented by the magnitude of the smallest seismic event that would be detected with a 90% probability by three stations or more. The *lower part* of the figure shows the estimated improvement over current capability from bringing the remaining 11 IMS primary seismic stations into operation. In these figures, array stations are shown as *filled circles*, whereas *filled triangles* denote three-component stations. One primary seismic station is missing from the calculations, as its location is not yet known (originally intended for India). The calculations in this figure are based on methodology described by Kværna and Ringdal (1999). Figure preparation courtesy of Tormod Kværna, NORSAR

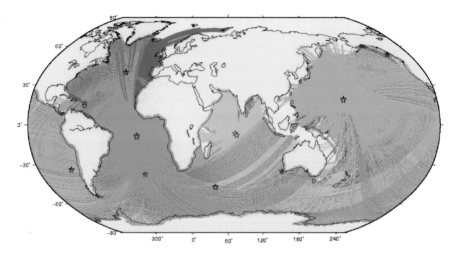

Fig. 7.3 Map showing the coverage of the world's oceans by the completed 11-station hydroacoustic component of the IMS. Colored ocean areas are "seen" by stations marked by that color. Many parts of the oceans are covered by more than one station, causing an overlap of colors. Near-coastal waters not covered by hydroacoustic stations are in many cases well monitored by seismic stations on land

although the screening conducted by the IDC will be of some assistance in the national interpretation effort, a large number of events will still require special attention by the designated national institutions.

The hydroacoustic component of the IMS has been completed and its coverage of the ocean areas is shown in Fig. 7.3. As can be seen in this figure, all major open ocean areas of the world are well covered. Explosive events in near-coastal waters that are topographically blocked from being observed at hydroacoustic stations, can in many cases be detected at onshore seismic stations. Signals detected at hydroacoustic stations are now contributing to the detection and location of events in oceanic regions together with the seismic data, in line with the expected synergies between the seismic and hydroacoustic networks. This implies that the detection capability of the combined seismic-hydroacoustic system will eventually be better in oceanic regions, especially in the southern hemisphere, than that shown in Fig. 7.2 for the seismic system. Recently, the PTS has demonstrated that the hydroacoustic stations that include triplets of hydrophones are efficient at recording not only hydroacoustic signals but also seismic ones. Improvements have been made to the processing algorithms to take advantage of this, which will contribute both to a further lowering of event detection thresholds and to improved event location accuracy. The hydroacoustic data have also provided some interesting scientific results; one such example is the IMS station at the Chagos Archipelago in the Indian Ocean, which shed considerable light on the time-history of the rupture process for the large Sumatra earthquake that caused the disastrous tsunami in the Indian Ocean on 26 December 2004; see Fig. 7.4. Research in the hydroacoustic

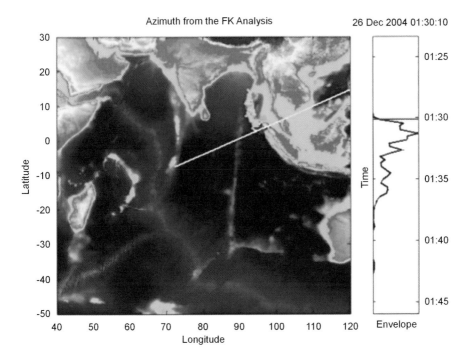

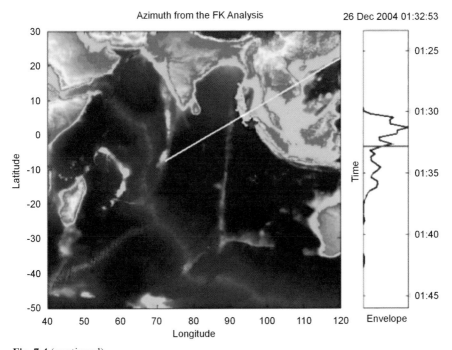

Fig. 7.4 (continued)

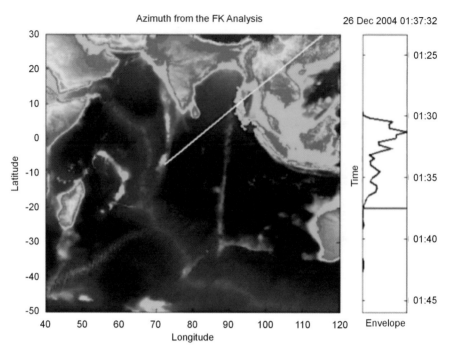

Fig. 7.4 Snapshots illustrating how the hydroacoustic IMS station at the Chagos Archipelago in the middle of the Indian Ocean over a time span of more than 7 min tracked the rupture process of the large Sumatra earthquake on 26 December 2004 that caused the disastrous tsunami in the Indian Ocean. The *red star* denotes the point of initiation of the earthquake rupture process. The panels to the right of the maps show the "envelope" of the hydroacoustic recording, and the *red lines* in this envelope indicate the times corresponding to the direction of approach of hydroacoustic waves arriving at the station, indicated by the *white lines* in the maps

community is progressing in areas such as the use of seasonally varying wave propagation models for event location, understanding hydroacoustic wave blockage by shallow ocean features and islands, and the effects of the SOFAR channel's disruption in Antarctic waters. Progress in these areas will help improve the performance of the IDC hydroacoustic data processing.

As we have described earlier, work on infrasound technology had been rather sparse in the last decades before the CTBT negotiations. As a consequence, efforts during the first ten years of the PrepCom have been spent, both inside the PTS and outside, on rather fundamental issues related to this technology. This work has included sensor development and station design, understanding of wind-generated noise and techniques for its reduction, study of infrasound propagation in a time-varying atmosphere, and development of algorithms for processing infrasound data. As data from the IMS infrasound network have become available, many special studies have been conducted into topics such as mapping of noise level variations, discrimination of earthquakes and mine blasts, and source location using

The results above refer to a network with 44% of all of the IMS radionuclide stations in place. The atmospheric transport modelling has until recently been carried out for a period of up to 6 days. With extended computer capacities at the PTS, this time period has now been extended to 14 days, which was also the period used in the modelling conducted to assess radionuclide network performance during the CTBT negotiations. With this consistency in the allowance for atmospheric transport times, the demonstrated high sensor sensitivity, and the full 80 station IMS radionuclide network in place, it is very likely that the capability will be very close to the projected performance used as a basis to guide the network design during the CTBT negotiations. It should also be noted that there is still considerable uncertainty in atmospheric modelling, which will be improved as our understanding of the physics of the atmosphere and its dynamics increases. Observations from the radionuclide network are likely to make significant contributions to the modelling of aerosol transport.

The corresponding capability estimates for the 40-station system to detect radioactive noble gases is not yet available. This system is expected to be a most important component, having capabilities to detect radioactive releases even from underground nuclear explosions, as discussed in Chapter 2.

7.2.4   Tests of the Monitoring System Provided by Nuclear and Other Explosions

The capabilities of the evolving IMS/IDC can also be assessed through consideration of the actual performance for events of special interest or significance, which have occurred during the build-up phase. The performance for several such events of relevance to the monitoring of the CTBT is described below.

At the time of the nuclear explosions conducted by India and Pakistan in May 1998, the IDC had not yet started its operations. In the context of the GSETT-3 experiment, the Prototype IDC (PIDC) in the USA, however, at the time of these explosions was receiving and processing data from a global network of 36 primary and 58 auxiliary seismic stations, the large majority of which were designated IMS stations. India announced that it had conducted nuclear explosions on 11 and 13 May, composed of series of three and two devices, respectively, whereas Pakistan announced the conduct of nuclear tests on 28 May with a series of five simultaneous explosions, and on 30 May as a single test. As described in Barker et al. (1998), the PIDC automatically detected and located the tests carried out on 11, 28 and 30 May and published the results in its Reviewed Event Bulletin within a few days. As the actual location of the tests on 11 and 28 May could be inferred from satellite imagery, it was possible to determine the true bulletin mislocations for these tests, which were less than 10 km in both cases. Barker et al. (1998) also demonstrated that location uncertainties can be further reduced using calibrated source locations and historical seismograms, so that the location uncertainty area for future events of similar magnitudes in this region will be of the order of 100 km^2 only. As to the size of these explosions, Barker et al. (1998) indicate explosive yields in the ranges 9–16 kt for the 11 May explosion, 6–13 kt for 28 May and 2–8 kt for 30 May. All these estimates are well below the yields announced by India and Pakistan, which were 55 kt, 30–35 kt and 12 kt, respectively. It was not possible to verify the number of individual charges in each explosion. The announced Indian test on 13 May was not detected by any GSETT-3 station, nor any other seismic stations which made data available to the international scientific community. Barker et al. (1998) concluded that an upper bound of the yield of this test was 30 tons, if it was performed under similar conditions as the 11 May explosion. Wallace (1998) provides additional, interesting assessments from interpretation of recordings of these explosions.

Large chemical explosions are valuable sources for the assessment of network capabilities. For example, on August 22, 1998, 100 tons of conventional explosives were detonated at the former Soviet Union test site at Degelen Mountain in Kazakhstan. The explosion was carried out by the Republic of Kazakhstan, in cooperation with the USA, to seal tunnels at this former test site. The PIDC's Reviewed Event Bulletin for this explosion showed a magnitude 3.8 event, which was detected at eight IMS-designated primary seismic stations, at distances up to 10 000 km. The PIDC's estimate of the location was approximately 12 km from the true location. Douglas et al. (1999), in a study of lessons learned from this explosion in the context of future CTBT monitoring, stated that "*The seismograms recorded are sufficient to identify the disturbance as suspicious and, if the test-ban treaty were in force, could have led to a demand for an on-site inspection to determine whether a nuclear test had taken place*". As we note later, a 12.5 ton explosion conducted in the same area as part of an OSI field experiment in 2002 was also well recorded by the IMS. In 1999, five chemical explosions were detonated in the Dead Sea for the purpose of calibrating seismic travel times in the Middle East and the Eastern Mediterranean to improve location accuracy of seismic events in the region, and to contribute to enhancing IMS capabilities in general (Gitterman and Shapira 2001). The largest of these explosions was a 5-ton shot, which was observed at eight IMS-designated seismic stations, located at distances up to 5000 km away from the source. The PIDC assigned a magnitude of 3.9 to this event, and the estimated location, based on IMS-designated stations, was only 2.3 km from the true location.

On 9 October 2006, North Korea announced that it had carried out an underground nuclear test that day. This was the first nuclear test carried out anywhere in the world since the IDC started operations in February 2000, with regular issuance of its Reviewed Event Bulletin. Whilst bearing in mind that the IMS and the IDC, in accordance with current PrepCom guidance, operated in a provisional mode, it is of considerable interest to assess how the current resources of the IMS/IDC responded to this event. Within 2 h of the event's origin time, an automatically determined initial event location, based on detection of signals at 13 IMS primary seismic stations, was available to States Signatories. During the interactive review phase over the next two days to produce the Reviewed Event

Bulletin for 9 October, the IDC analysts added information derived from one additional primary seismic station and eight auxiliary seismic stations to form the event solution for the North Korean test. The event was given a magnitude of 4.1 in the bulletin, and the size of the location uncertainty area was 880 km^2 (see Fig. 7.9). Kim and Richards (2007) show the presumed actual location of the test, inferred from satellite imagery, and this location is well positioned within the uncertainty area of the IDC's location.

In the absence of an announcement by North Korea, would it have been possible for States Signatories to identify this event as a nuclear explosion? In other words, would they be in a position to independently confirm the North Korean announcement that a nuclear test had taken place, using IMS and other data available to them? In accordance with its assigned role, the PTS is not making explicit statements on the nature of events in their IDC products distributed to States Signatories. As we described above, however, all events in the Reviewed Event Bulletin are subjected to screening, and the North Korean test was *not* screened out as earthquake-like. What else is it possible to say about the source, based on seismic recordings alone? Using data from IMS seismic stations as well as data available from other seismic stations, members of the scientific community have concluded that the data are indicative of an explosion rather than an earthquake. For example, Richards and Kim (2007) have studied seismic data for both earthquakes and chemical explosions recorded at a station in China located at a distance of 370 km from the North Korean test site to demonstrate that this station's records of the North Korean test clearly fall into those of the explosion population. But it is still not possible to discriminate on the basis of seismic data alone between a nuclear explosion and a chemical one in which all explosive material is detonated simultaneously. Moreover, chemical explosions of magnitude 4, with simultaneous release of all energy are feasible, and have been carried out in the past. So what other data are available to States Signatories to answer the question: Was this explosion nuclear or chemical?

As we have noted, the IMS radionuclide network will have equipment for noble gas detection at 40 of its 80 station locations upon entry into force of the Treaty. Radioactive noble gases are, as we have described earlier, of special relevance to underground nuclear explosions. These gases are the ones carrying radionuclide products that would leak most readily to the surface and be discharged into the atmosphere, for an explosive device that is well contained. At the time of the nuclear test by North Korea, ten noble gas stations, capable of detecting four xenon isotopes and metastable states, were active in the International Noble Gas Experiment, undertaken by the PTS. As already described in Chapter 2, one of these stations, located at Yellowknife in Canada, detected Xe-133 on 21 October 2006 and also on several days afterwards, at levels above the expected Xe-133 background on those days (Saey et al. 2007). As shown in Fig. 2.12, forward atmospheric modelling of a possible release of Xe-133 from the event located by seismic methods in North Korea on 9 October 2006 showed that such a release most likely would have been

Fig. 7.9 Illustration showing the location of the North Korean nuclear test on 9 October 2006 as calculated by the IDC from IMS seismic data, as well as associated uncertainty areas. The *larger, blue confidence ellipse* is associated with the automatic data processing, whereas the *smaller, red ellipse* resulted from the subsequent interactive review by the IDC analysts. The area inside the *red ellipse* is 880 square kilometers, which is less than the maximum allowed area of 1000 square kilometers for an on-site inspection under the Treaty. The seismic recordings are from the IMS primary seismic station PS31 in South Korea. The recording shown in *red* is from the North Korean test, whereas the blue trace is from an earthquake in 2002, which occurred not far from the test site. Note the differences in the characters of the two recordings

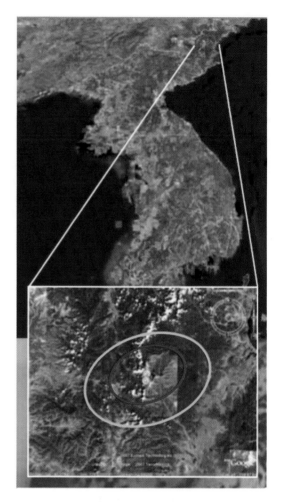

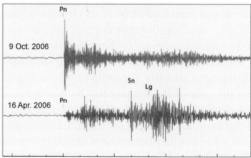

transported to the area of Yellowknife by October 21. Atmospheric modelling in the form of backtracking has also shown that this detection of Xe-133 appears to be consistent with a possible release of noble gases from the North Korean event. Radionuclide observations of relevance have also been reported by non-IMS monitoring resources: on October 16, 2006 the U.S. Office of the Director of National Intelligence reported detection of radioactive debris in air samples collected on October 11, 2006 "*which confirms that North Korea conducted an underground nuclear explosion in the vicinity of P'unggye on October 9, 2006*" (Office of the Director of Intelligence 2006). Furthermore, as also noted in Chapter 2, the Swedish Defence Research Agency (FOI) announced that a mobile system for detecting radioactive xenon had been deployed in South Korea and started collection of air samples within three days of the event in North Korea (Fig. 7.10). All samples collected

Fig. 7.10 Lars-Erik De Geer of the Swedish Defence Research Agency (FOI) with the SAUNA system developed at FOI for very sensitive radioxenon detection. At the end of 2007 seven such systems were installed in the IMS network. Immediately after the nuclear test in North Korea on October 9, 2006 a mobile version was flown by FOI to South Korea, where several xenon signatures indicative of the test were detected

were found to contain radioactive xenon, and FOI concluded, from atmospheric transport modelling, that the xenon, with a high level of probability,
originated from the area in North Korea where the explosion took place
(FOI 2006, Ringbom et al. 2007).

7.2.5 Assessment of Eventual IMS/IDC Capabilities; Room for Further Improvements

We have seen that the currently projected overall event detection capability of
the seismic component of the IMS matches the expectations at the time of the
CTBT negotiations. There is a potential to exceed these expectations, as favorable receiving conditions in terms of low background noise have been observed
or can be expected at stations yet to be built. There is also a potential for
improvements to station signal detection processing. We have seen that the
IDC has demonstrated an ability to issue automatic and reviewed event bulletins of high quality in a timely manner for events of special interest, like the
unplanned test of the system provided by the North Korean nuclear explosion.
Moreover, there is good reason to believe that the IDC will further improve the
accuracy of seismic event location through ongoing station tuning and wave
propagation calibration efforts.

   We have so far focused on the global detection capability of the IMS seismic
network by presenting global maps showing detection thresholds, which are
estimates of the *smallest* magnitude of seismic events that could possibly be
detected. As we have explained, detection is usually required by a minimum of
three stations to provide a meaningful location of the event. This is the traditional way of assessing network capability, and calculations performed to
derive the capabilities shown in Figs. 7.1 and 7.2 are based on the assumption
of detection at three stations. The IMS three-station detection capability at the
90% probability level for the North Korean test site at the time of the North
Korean nuclear explosion was at approximately magnitude 3.8. One might say
that the requirement of detection at *three* stations or more is too conservative,
or stringent, in the sense that its application does not show the full potential of
the network. Ringdal and Kværna (1989, 1992) have developed a technique,
referred to as "threshold monitoring", which represents a supplement to traditional event detection analysis and in which they address the question: Given
the data available, what is the *largest* seismic event at a given site or in a given
region that could possibly have *occurred*? By subjecting the North Korean test
site to threshold monitoring, Kværna et al. (2007) conclude that this site, using
data from the IMS network, can be monitored, in the threshold monitoring
sense at the 90% probability level, down to a magnitude of between 2.3 and 2.5,
corresponding to explosive yields of only 10 tons or so for explosions in hard
rock. Not surprisingly, a seismic array station in South Korea is essential to
obtain such low values. The threshold monitoring approach provides an upper

limit of the magnitude of events that have *not* been detected, and is useful for identification of times, or intervals of time, when the possibility of "hidden" seismic events is particularly high.

The main purpose of the threshold monitoring method is thus to draw attention to those time instances when a given threshold (e.g., 2.3–2.5 for the North Korean test site at the 90% probability level) is, in fact, exceeded. For such instances, the analyst will then apply other, traditional analysis tools to clarify the source of the observed disturbance and if possible, confirm that there was in fact an event at the time and location in question. The threshold monitoring technique can be said to provide a fuller picture of actual capabilities than the traditional approach based on detection at three stations, and in that sense represents an optimal usage of available resources in terms of sensitive IMS stations. This technique can be useful to States Signatories that might find reason to subject specific areas of interest to special scrutiny, continuously or during limited periods of time. The substantial lowering of the network's threshold, in the sense explained here, could represent a considerable deterrence to clandestine testing under the Treaty.

Both the three-station detection capability and the threshold monitoring capability are commonly estimated at the 90% probability level, to provide high confidence in the results. As we discussed in Chapter 2, a potential cheater would most likely not take more than a 5–10% risk of being detected. So our potential cheater would, from his own perspective, face capabilities that would be clearly better than those presented in this chapter for the seismic component of the IMS. For example, at the 5–10% probability level, the three-station detection capability would be of the order of half a magnitude unit lower than the capability corresponding to the 90% probability level.

The IMS hydroacoustic stations have clearly shown their potential, especially in synergy with IMS seismic stations, and are contributing to the definition of events, not least in the southern hemisphere, as intended. The IMS infrasound stations have demonstrated their ability to detect signals from atmospheric phenomena, but as network processing has just started it is too early to reliably assess the eventual global capabilities of the emerging infrasound network. The IMS radionuclide network is at an advanced stage for the particulate stations; relevant spectra are subjected to interactive analysis and products of atmospheric transport modelling are generated on a regular basis. The International Noble Gas Experiment has, as we have seen, contributed significantly to the development of the noble gas monitoring technology, and the North Korean event clearly demonstrated the potential of noble gas monitoring in the CTBT context. Following this event, PrepCom's Policy Making Organs agreed to give the noble gas technology higher priority, and as we saw in Chapter 6, a plan has been developed to move the noble gas detection systems into provisional operation. Results and data from noble gas stations are now also being made regularly available to States Signatories.

Work needs to progress towards further enhancements of IDC data processing capabilities. In Chapter 2 we pointed out the dramatic development in data analysis and data management, often referred to as data mining, and possible applications to enhance the processing of the seismic data at the IDC, in particular. As mentioned there, one novel approach in this regard is offered by so-called cross-correlation analysis, which utilizes entire waveforms and not just a few extracted parameters. The idea is to quickly and automatically recognize new seismic events that essentially are realizations of events that have occurred in the past. This is achieved by correlating, in real time, the incoming continuous data stream with waveforms recorded for past events. It has been shown that if a cross-correlation value calculated in this way exceeds a certain level, the two events in question essentially originate from the same location and are of a similar nature. Figure 7.11 shows the application of this technique to identify a small event that is completely masked by a much larger event that occurred at the same location some 45 s earlier. The IDC has done some limited experiments with this technique to identify aftershocks, some thousands of which sometimes follow major earthquakes. Successful

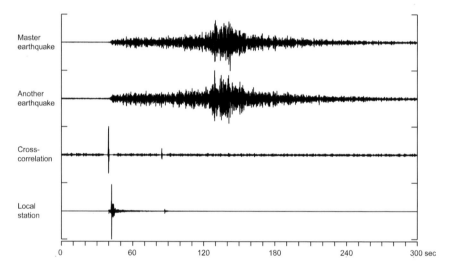

Fig. 7.11 Figure illustrating the potential of the waveform cross-correlation method to detect signals buried in the recording of a much larger event, and to identify these buried signals as coming from an event very similar to the larger one, both in terms of location and nature of the event. The trace at the *top* of the figure ("Master earthquake") is the recording of an earthquake in northern Norway made at an IMS station in southern Norway at a distance of 600 km. Trace number two from the *top* is a recording made at this IMS station of another earthquake at the same location in northern Norway. Trace number two from the *bottom* is the cross-correlation of these two recordings, and it shows that the two waveforms correlate well, as indicated by the first "spike" in this trace, implying that the two earthquakes are similar in nature and that they occurred at the same location. The cross-correlation trace has a second peak, however, significantly above the background level, showing that there is another event – an aftershock – with signals buried in those of the main shock, which then also correlates with the master recording. The *bottom trace* is a recording from a station located only some 15 km from these earthquakes. This station independently confirmed this small aftershock some 45 s after the main shock. Figure preparation courtesy of Steven Gibbons, NORSAR

application of the technique to aftershock sequences would save considerable time and resources at the IDC, but the ambition should be to test and apply this technique more generally. The technique could be used as a first pass of the data to screen out those events that match previous events, which are already well determined and well understood. Events that are not screened out in this way could then be subjected to traditional analysis and sometimes special scrutiny, as they would have no match in previous history.

In this chapter we have described the capabilities and potentials of the individual technologies that make up the IMS, and some synergies between them. At the end of the day, however, it is the combined resources of the IMS networks, the ability of the IDC to process the data and issue their products, and then the capacity of individual States Signatories to analyze and assess the data and products that will prove decisive. From performance already demonstrated by the IMS and the IDC in their current, advanced stages of development as well as current projections of the capabilities of the completed systems, there is every reason to believe that the monitoring component of the verification regime will represent considerable deterrence to any clandestine future testing of nuclear weapons.

The last link in this verification chain, then, is the analysis and assessment by States, and we discuss this further in Chapter 8. We are concerned that this might be the weakest link, as many States have substantially reduced their competence and resources in this area. Today only few States are able to carry out an independent analysis and assessment of observed events for Treaty verification purposes. We do not think this ability should be limited to a few States, but rather that States around the world should be able to form their own, independent opinion on observations provided by the IMS/IDC. If they are not to be dependent on large States, State Signatories around the world will have to establish the necessary competence and resources, either nationally or together with neighboring States in regional cooperation. National assessment is a most essential element in the verification chain and States need to take the responsibility to improve on this capability worldwide.

7.3 Exercises Also Get OSI on the Move

As commented in Chapter 6, it was initially a difficult task to effectively organize the work in Working Group B related to OSI issues. The preparation for, conduct of and evaluation of a large-scale experiment in Kazakhstan in September–October 2002, however, changed this picture and injected new enthusiasm and vigor into the process, and the results of this experiment gave very useful directions for further work.

The conduct and outcome of this field experiment over a 550 km^2 inspection area involved 27 surrogate inspectors from 17 States Signatories and the PTS. The inspectors spent two weeks in the field, performing activities much like those of a real inspection team, subsequent to entry into force of the Treaty.

More than two tons of equipment was shipped from Vienna to Kazakhstan. The experiment began with the simulation of an underground nuclear explosion by detonating 12.5 tons of chemical explosives at a depth of 200 m in an unused borehole at the former Soviet Union test site at Semipalatinsk. The explosion was well recorded by the IMS. The location calculated by the IDC and its associated uncertainty area were not, however, disclosed to the inspection team because this location was regarded as too accurate, and instead a fictitious location and uncertainty area were provided. The inspectors employed a range of measurement techniques, integrated together for the first time to ascertain synergies among them. The techniques employed included use of portable seismometers and in-field analysis of the data collected, collection of air and soil samples to search for OSI-relevant radionuclides, and visual inspection, including low altitude helicopter overflight, in search of indications of recent human activity. The crater formed by the explosion was filled in and rehabilitated so the landscape was similar to its appearance before the explosion. The "ground zero" for the experiment still had some visual differences from its surrounding area, but there were many other visual anomalies spread across the inspection area. Small aftershocks, which often follow real nuclear tests, were simulated by kilogram-sized chemical explosions, which were conducted without informing the inspection team, to increase the realism of the inspection exercise. The inspection team did not locate ground zero. It turned out that the only events that could have been directly useful to the inspection team in finding ground zero were some small seismic events that were detected at the end of the exercise by one of the participating seismic experts, resulting most likely from settling of the debris that was used to fill up the crater (including parts of trucks). This finding was, however, not consolidated by the team as a whole. Subsequent analysis showed that these seismic events originated exactly from the site of the crater that had been covered. An important lesson learned from this was that available data during an inspection will be rather sparse and that it is important to thoroughly check every bit of data before discarding it. Six evaluators observed every aspect of this field experiment and recorded a large number of lessons learned. Based on these observations, other observations by participants in the experiment and its subsequent evaluation by the PTS, the experience was condensed into approximately 140 lessons learned, giving rise to some 300 implementable actions.

The PTS efforts related to the conduct of the 2002 field experiment and its evaluation gained widespread recognition among States Signatories. No doubt, this event served to direct much needed attention to the OSI program, which was lagging behind the other verification-related programs in terms of level of development and maturity. The external evaluation of the OSI program conducted in 2003 (CTBT-WGB-21/INF.5 2003), emphasized the value of the 2002 experiment and suggested the conduct of another major exercise. This evaluation team also proposed that the PTS should develop a strategic plan for the continued build-up of the OSI regime. The PTS embarked on the development of such a plan and after a few rounds of reviews and revisions, taking account of the prospects for allocation of future resources, a two-phase plan was presented in 2005. The first phase will establish provisional capabilities through the development, testing and refinement of procedures and tools needed for the eventual conduct of OSIs. In this regard, field exercises are essential to gain practical experience, and will be conducted periodically during this phase. The second phase would start fairly close to entry into force of the Treaty, and would be devoted to a rapid preparation of the OSI regime, including additional training of inspectors and procurement of equipment.

In 2005, the Policy Making Organs approved the PTS plan for the conduct of the next major exercise, to be conducted in 2008. This exercise will also take place within the former nuclear test site near Semipalatinsk in Kazakhstan. The PTS, the Policy Making Organs of the PrepCom as well as States Signatories are now focusing their efforts on this field exercise. Based on the ongoing work in WGB to develop the OSI Operational Manual, an OSI Test Manual has been completed to guide the conduct of the exercise in 2008, as well as PTS activities and related training leading up to this exercise. As things currently stand, the 2008 exercise with all its planned activities including evaluation, promises to be another boost for the development of the OSI regime, and hopefully, together with smaller directed exercises, will go a very long way towards demonstrating operational readiness of the OSI regime.

The OSI work so far, in particular the field tests conducted and the one planned for 2008, has focused on activities to detect, locate and identify a clandestine test. An OSI might, maybe more likely, be conducted in a situation where no clandestine test has been conducted, but where concern was about a naturally occurring event, such as an earthquake. How much inspection is enough? When and how can you finish an OSI and leave the scene with increased confidence, ever when you have found nothing? It is essential that, in particular, State Signatories also discuss and develop a strategy for such situations.

Chapter 8
National Technical Implementation of the CTBT

8.1 National Commitments

Institutions in States Signatories play crucial roles in the implementation of the CTBT. In this chapter we first look at various activities at these institutions, and then provide an assessment of the status of such activities ten years into the PrepCom phase.

In order to fulfil its obligations under the Treaty, as noted in Chapter 4 each State Party shall designate a National Authority which will serve as the focal point for contact with the CTBTO and other States Parties. By October 1, 2007, 134 States Signatories had notified the PrepCom of their designation of National Authorities or 'national focal points'. A National Authority is typically a function accommodated by an office of the ministry of foreign affairs, or another ministry of the State Signatory.

The National Authority normally needs to be supported by an institution that can handle all technical aspects related to the State's participation in the verification-related activities. This function, accommodated by some technical institution, is referred to as the National Data Center (NDC). This is not an entity in the Treaty, with well-defined functions; it appears only in lower case letters (national data centers) in the context of one way of routing of IMS data to the IDC, and in the context of receipt and analysis of IMS data. Many States have established NDCs; Fig. 8.1 is a picture of the NDC established by China in Beijing. There are two main dimensions to the activities at NDCs and other designated technical bodies in States Signatories:

1) To ensure that States Signatories fulfil their obligations in establishing IMS facilities on their territory, in provisionally operating and maintaining such facilities, and in facilitating the provision of data from IMS stations on their territory via the Global Communications Infrastructure to the IDC. NDCs may also take part, on a voluntary basis, in work to further develop and improve the various elements of the verification system; and
2) after entry into force of the Treaty, to provide interpretation and expert advice to National Authorities on events reported by the CTBTO, as a basis for decisions by those authorities on issues related to compliance with the Treaty.

O. Dahlman et al., *Nuclear Test Ban*, DOI 10.1007/978-1-4020-6885-0_8,
© Springer Science+Business Media B.V. 2009

Fig. 8.1 The newly established premises for China's National Data Center in Beijing

During the PrepCom phase, the tasks of these national institutions are to support the activities under (1) above, and to make necessary preparations to be able to conduct activities under (2) at the appropriate time.

8.2 National Institutions Play Key Roles in the Global Verification System

The host States are obligated to cooperate with the PTS in the establishment of IMS facilities and necessary means of communication, and are expected to take responsibility for the provisional operation of these facilities, as well as for transmission of data to the IDC. To carry out these activities, the hosting States sometimes designate national institutions with appropriate competence, often an already existing NDC, to negotiate contracts with the PTS for execution of these tasks. At other times such contracts are awarded by the PTS following a competitive bidding process. Either way, the PTS relies on competence in States Signatories to carry out essential work for the establishment and proper functioning of the monitoring component of the verification regime. Such competence is available in several States which have a history of establishing and operating stations, in particular seismic ones, that later became IMS facilities. Other States lack such experience, and the PTS has had a rigorous program over the years to train operators of IMS stations. To date, about 600 trainees from around the world have participated in such courses. With the exception of auxiliary seismic stations, all costs of operating certified IMS stations should be covered, in principle, through contracts let by the PTS. To a large degree, too, the PTS is footing the bill for direct operational expenses and manpower needed for the immediate and concrete operational tasks. It is, however, proving difficult for the PTS to financially support the broader institutional competence required to secure quality assured operation of IMS stations around the world, even for the current provisional mode of operation. Therefore there is a need for solid commitments on the part of States hosting IMS stations to take on the responsibility for maintaining national competence and associated infrastructure, to ensure the proper functioning of the global monitoring system in the longer term.

There are important tasks for national institutions in the evaluation and improvement of the quality of IDC products. On a voluntary basis, a number of NDCs and other national institutions that regularly receive bulletins and other products from the IDC provide feedback on these products to the IDC, based on their own analysis and knowledge. This might include information on the true position of mine blasts and other events appearing in the IDC bulletins. Such feedback is indispensable to enhancing the performance of the global monitoring system. Another example of a successful activity is the engagement by many NDCs in evaluation of the system-wide performance test in 2004 and 2005. In fact, this test helped to establish the important role of NDCs in the development of the verification system and fostered good cooperation between the NDCs and the PTS. One outcome of this testing has been that NDCs' engagement in evaluation activities has increased significantly. The NDCs were also instrumental in defining follow-up, focused tests of various elements of the verification system over the next few years. In general, NDCs' participation in such activities will increase the credibility of and the confidence in the verification arrangements.

Efforts by institutions in States Signatories are important in other contexts, too. The PTS issues contracts, following competitive bidding, for a range of activities related to the development and further enhancement of capabilities of the IMS, IDC and OSI elements of the verification regime. In this way, specific knowledge in national institutions is exploited to the benefit of the system. But expertise residing in NDCs and other national institutions is also made available to the PrepCom free of charge through participation in Working Group B and related activities like technical workshops. Any country's participation in WGB has two purposes: to promote the specific interests of that country and to contribute to the advancement of the verification regime at large. It is fair to say that the main activities in WGB are devoted to cooperative efforts to find technically sound solutions to issues at hand, and in this respect experts contribute their knowledge in good faith and to the benefit of the common good. So how many States are currently active in WGB at a technical level? A survey of participation in WGB sessions over the last year indicates that one or more delegates of 30–35 States are affiliated with technical institutions in their home countries. This is not an impressive number, considering the current number of

178 States Signatories. The number of technical experts representing developing countries is low. These countries have argued that, for economic reasons, they are largely unable to send their technical experts to WGB sessions. To remedy this situation, effective from 2007 voluntary funding has been provided by some States Signatories to finance participation in WGB sessions by experts from developing countries.

8.3 Basis for National Interpretation

As pointed out in several contexts in this book, matters related to possible non-compliance with the CTBT must be raised by States Parties themselves. Raising such matters will involve considerations that are likely to be based both on the interpretation of available verification data, and circumstances of a political nature. This implies that the National Authorities need to be supported by technically competent institutions that have the ability to interpret data and products and alert their authorities to events of concern. This is thus a national matter and each State must decide for itself what level of in-country expertise it requires or what expert support it may acquire in cooperation with other States.

In their efforts to identify events and radionuclide detections of possible concern after entry into force of the Treaty, States Parties will be supported by the Technical Secretariat of the CTBTO in several ways. In the products of the IDC, events detected by the IMS stations are objectively characterized in the form of parameters (values for certain, measured physical properties) that provide indications to its recipients about the nature of these events. As explained in Chapter 4, the IDC will also produce so-called screened bulletins. To create these bulletins, events that are likely to be natural or man-made, non-nuclear are removed, on the basis of certain criteria. The idea is to allow the States to focus their attention on the remaining events. Screened bulletins may also be produced on request by any State Party, based on criteria provided by that State. And any State Party is entitled to all IMS data and copies of all software used by the IDC in its processing of IMS data, so that States will have the opportunity to process any IMS data themselves, for whatever reasons they may have.

It is the States Signatories themselves that must take on the responsibility during the PrepCom phase to prepare for their role after EIF with respect to national interpretation. One indicator of the level of activity in this regard is the States Signatories' current use of IDC services in their preparatory work. In order to take full advantage of current IDC activities, designated personnel in States Signatories must sign up for secure signatory accounts, which enable access to IDC databases containing IMS data and IDC products. By the end of 2007, 859 users in a total of 97 States Signatories had established such accounts. In 2006, about 1.6 million IDC products and data segments/frames from IMS stations were sent to users in national institutions, on the basis of regular subscriptions or in response to specific requests. Of these, about 448 000 were bulletins produced from analysis of mainly seismic, but also hydroacoustic and infrasound data and 149 000 were products derived from radionuclide data. Among the other data distributed to States Signatories were about 150 000 segments of data from seismic, hydroacoustic and infrasound stations, 210 000 messages containing radionuclide data, and 327 000 messages with meteorological data. Moreover, nearly 2000 gigabytes of continuous data from seismic, hydroacoustic and infrasound stations were supplied to NDCs in 2006. The PTS offers to States Signatories a software package containing some basic data processing and analysis utilities for seismic, hydroacoustic and infrasound data, and currently institutions in approximately 100 States Signatories have received this software. The data, products and the software received from the PTS are also useful to States in other contexts, such as the analysis of earthquakes of special interest, and this may even be the main motivation for some States to obtain access to these services.

The development of competence in the area of event identification is crucial if NDCs are to fill their national role. This is an area where research is going on internationally, with findings being openly published in the scientific literature, with additional, classified research being carried out in some countries. There is an extensive, open literature on methods to discriminate between earthquakes and explosions, based on the analysis of seismic data. NDC personnel need to stay abreast of the current status of this research to be able to serve their authorities well. Where do matters stand with respect to event identification, and what are the challenges to those involved nationally to advise their authorities? We have presented thresholds for event *detection* in this book, for example in terms of magnitudes of seismic events. The threshold for seismic event *identification* when recordings are made at distances exceeding 2 000–3 000 km, is believed generally to be about one half magnitude unit above the detection threshold. For recordings made at closer distances, practical experience is more limited, but the ability to discriminate between earthquakes and explosions has been demonstrated down to magnitude 3 for many regions for which high quality data are available (NAS 2002). For the IMS network, the event detection threshold is determined solely by the 50 primary seismic stations. For event identification, the additional 120 auxiliary seismic stations are available, and there is thus reason to believe that the event identification threshold of the entire IMS seismic network is not much higher than its event detection threshold. Synergies between the various IMS technologies are important in event identification. For example, events confidently located by seismic stations in oceanic areas are very likely to be earthquakes, unless strong hydroacoustic signals are present, indicating an explosion in the water masses.

The expertise required to provide national interpretation would often be part of an NDC or another institution that is also handling the mandatory and voluntary tasks described in the previous section. The competence needed for national verification purposes can thus often also be used to promote general improvements to the global verification system, to the benefit of all States.

Employees of such institutions often have a broad scope to their activities, and a given research project may build national competence while at the same time benefiting the global system, for example in the form of calibration of the IMS network, which will improve on IDC products. There is a culture of international cooperation and openness among researchers from the various NDCs, resulting in the sharing of their research findings. Some NDC environments also have researchers who respond to various solicitations for unclassified nuclear monitoring related research. Such contract research, based on competition in the scientific community, is supported, for example, by several US government agencies.

A request by a State Party for an on-site inspection shall be based on IMS data, on relevant technical information obtained by national technical means of verification, or on a combination of these. This implies that the designated national institutions must develop and maintain competence also to handle relevant national technical means, should such means be available, and to use such information when concerns of non-compliance occur. For example, non-IMS seismic stations, of which there are thousands around the world, may shed light on events of concern detected by the IMS. An NDC that has access to such data, and relevant data from other technologies, is in a good position to advise its National Authority. The national tasks are thus multifaceted, and an NDC function anchored in an institution with the widest possible expertise relative to CTBT relevant issues will be a considerable asset to any State Party.

National institutions working in S&T fields related to CTBT verification are essential in a national perspective to support decision making authorities. Work at such institutions is also essential to underpin the long term viability of the international verification regime. In this context it is a matter of concern that many States have downsized their national technical and scientific resources. Few States today have research organizations working in CTBT related fields or institutions able to interpret observations from all verification technologies. Moreover, the close international cooperation among national institutions that existed before the Treaty was negotiated has declined. It is essential that States realize that the CTBTO will not be responsible for verifying compliance with the CTBT; it will only be a service organization for States Parties, providing important information, which States then must have the ability to interpret. Once gone, it takes a considerable amount of time to re-establish this ability and the competence required at the national level to interpret CTBT verification data.

8.4 Regional Cooperation

It is understandable that several small countries and also many developing countries, for lack of resources, hesitate to take on the burden of establishing and developing NDCs for CTBT purposes, in particular if they lack an existing institution working in any of the S&T areas related to CTBT verification. The idea of regional centers has been around for quite a while. Groups of countries, perhaps typically within a certain geographical region, may see advantages in joining forces in the development of their NDC functions. Using modern technologies, including available communications solutions, tasks and responsibilities can be shared among participating countries, with a resulting cost saving for each of the States. All functions described above for individual NDCs can be accommodated by regional data centers. Such a regional center might be a physical center in the region, and maybe part of an already existing regional organization. It might also be a virtual center, connecting activities in several institutions in a region. The more sensitive political interpretation could be done at a national level, based on joint technical analysis at a regional center. When States participating in regional centers are also close allies politically, they may even want to make joint political interpretations and take joint action.

Such regional data centers could serve as focal points for training activities by hosting PTS training courses or training offered by States Signatories on a voluntary basis, and by conducting their own training to enlarge regional participation in the centers' activities. It could also be envisaged that such centers, after gaining the necessary skills, could take on some functions on behalf of the IDC. One idea that could be pursued is to let such centers, perhaps on rotation, participate in the stage of interactive analysis of IMS data before the issuance of the Reviewed Event Bulletins and Reviewed Radionuclide Reports by the IDC. By involving centers in different parts of the world, manned operations around the clock would be possible. Such farming out of real and essential functions on behalf of all States Signatories would no doubt contribute to further increasing competence and engagement around the world.

In Chapter 10 we point to a number of areas of civil and scientific applications of IMS data and IDC products. Regional data centers could take a key role in utilizing and exploiting IMS data, IDC products, software provided by the PTS, as well as various skills and knowledge gained in the CTBT context, for such purposes, to the benefit of their regions. These resources, used together with other available data, could create activities and address matters of importance to the regions, beyond monitoring the CTBT.

One key question is that of funding the establishment and operation of regional data centers. For resourceful countries, such funding should come from the pooling of resources available in the participating States themselves. For developing countries, funding for regional centers could be made available by donor states, or groups of donor states. Some States have already spent money on similar activities, mostly in bilateral undertakings, based on resources available in national budgets for promotion of capacity-building in developing countries in relation to nuclear disarmament and non-proliferation. The challenge in this regard is to make sure that the donors continue their assistance at least over the time it takes to develop the regional centers to the level that they become operational and are able to sustain the activities on their own.

Regional data centers for CTBT-related matters would be particularly useful in developing countries, both for these countries themselves but also as a means to broaden the participation in CTBT-related work. Despite widespread recognition of the potential benefits of such centers and intentions expressed at numerous regional workshops over the years, no such centers have yet been established. What is needed appears to be some well-focused, forceful donor countries with sufficient resources to patiently sustain an activity over some time, which are able to see the benefits to all States from helping to build capacity in developing countries and thus enlarging the number of States Signatories engaged in verification-related activities. Such a broadening of participation will anchor the verification arrangements in all corners of the world and will contribute to building confidence, to the benefit of all States Signatories.

The concept of regional centers may be of interest not only for developing countries but also in other regions. A joint EU center might be an interesting concept. The EU is strengthening its cooperation on security issues, and treaty verification might prove to be an interesting issue for such cooperation. It might also be an area where a joint institution, real or virtual, might prove realistic, especially since many EU member States have reduced their national competence and resources in S&T areas of relevance to CTBT verification.

8.5 An Assessment of the Status of National Technical Implementation

In this chapter we have described various roles and functions of national institutions in the implementation of the verification regime of the CTBT, and we have included some numbers of countries involved in various activities, to indicate current engagement levels. In our assessment the PTS has done a good job, within its mandate, to provide assistance to National Authorities, NDCs and other institutions in States Signatories, in order to inspire or initiate national activities. PTS has arranged workshops and general and specialized training courses, provided special advice in bilateral contacts, distributed information material and been proactive in contact with representatives of States Signatories at all levels. How well have States Signatories responded to these initiatives by the PTS as well as bilateral initiatives by other States Signatories? And what have the States done themselves so far during the PrepCom phase to contribute to the common efforts of the development of the verification regime and to prepare for their own national functions after entry into force of the Treaty? Given the current number of 178 States Signatories, it might appear a reasonable level of activity that institutions in 97 countries have as of late 2007 registered to receive data and products from the IDC. Looking behind these and other figures for nominal participation, however, our assessment is that after 11 years of the PrepCom phase, engagement by States Signatories is a little less than we had hoped for. Of the 94 States that had authorized their national institutions to access IMS data and IDC products by the end of 2006, only 37 obtained such data and products on a daily basis, whereas 18 States did not access the databases at all during that year. On a positive note, all six geographical regions were represented among the 37 States, with institutions in three States or more in each region receiving data and products on a daily basis. As to voluntary participation in the further development of the monitoring system, only about 20 NDCs can currently be rated as contributing substantially to this effort.

In their mandatory role to support the PTS in the establishment of the IMS, States have in general worked hard to make sure that stations can be built and communications channels installed in accordance with annual PrepCom program and budget decisions. But as we have seen in Chapter 6, national legal procedures and environmental concerns have sometimes delayed work substantially, and States are sometimes taking a long time to respond to the PTS in various stages of contract negotiation. States' engagement in contributing, on a voluntary basis, to the enhancement of the performance of the verification system could have been stronger and more comprehensive, as we have pointed out. The number of countries taking an active part in work to evaluate current performance and contribute to its improvement, for example, is lower than the number of States that participated in the activities of the GSE prior to the CTBT negotiations. We also note that several States that participated actively in the GSE are not sending technical experts to Working Group B's sessions. Nevertheless, there are some encouraging signs: the PTS managed to engage more NDCs in the evaluation of the system-wide performance test in 2004/2005 than in previous evaluation activities. And States are very enthusiastic about the OSI field exercise planned for 2008, and they are doing their best to support the PTS in its preparation for this exercise. As to national interpretation, it is our general impression that States are generally taking their time and are not moving very rapidly to build the necessary competence and establish the relevant functions, even though most recognize that sustained activity over a long period of time is needed. Many of them are perhaps delaying this work until the Treaty's entry into force is in sight. Yet others perhaps expect to be able to rely on advice from political allies when it comes to their assessment of Treaty compliance by other States.

Chapter 9
The CTBTO Preparatory Commission and the PTS – an Organizational Perspective

In this chapter we look at the CTBTO Preparatory Commission and the PTS from an organizational perspective, summarizing our experience and the results of the different reviews carried out of the PTS and its activities. Turning the spotlight closely on any organization brings out a number of things that can be improved upon and this case is no exception. This does not, in our view, mean that the PrepCom or the PTS have more problems or are less efficient than one might expect of any international organization. On the contrary, we want to recognize the demanding work carried out in Vienna and around the world to implement the CTBT verification regime and prepare for the entry into force of the Treaty. Our intent is not to criticize but rather to identify lessons to be learned. Lessons that may be used to improve the operation of the PTS and lessons relevant for the Technical Secretariat, to be established after entry into force. Some of this experience might also be useful when creating efficient organizations for treaties that might be negotiated in the future.

9.1 The Policy Making Organs

The purpose of the CTBTO Preparatory Commission is to carry out "*the necessary preparations for the effective implementation of the CTBT*". When analyzing the PrepCom and the PTS from an organizational perspective we should thus remember that the PrepCom was tasked with preparing for a future permanent organization, the CTBTO, and was expected to last for a limited time only. Some of the issues that have become hurdles in the now long life of the PrepCom and the PTS might not have appeared had the Treaty entered into force as originally expected within a few years.

9.1.1 The Structure of the Policy Making Organs

The PrepCom is open to all States Signatories, 178 of them as of February 2008, and some 100 States normally participate in its plenary sessions and those of its

O. Dahlman et al., *Nuclear Test Ban*, DOI 10.1007/978-1-4020-6885-0_9,   183
© Springer Science+Business Media B.V. 2009

Fig. 9.1 Representatives of States Signatories assembled in Vienna for a plenary session of the PrepCom

two Working Groups, Working Group A (WGA) for budgetary and administrative issues and Working Group B (WGB) for verification related issues (Fig. 9.1). In addition, the PrepCom has an Advisory Group composed initially of 12 experts to provide advice, in particular to WGA, on financial and administrative issues.

During its first years the PrepCom had three yearly plenary sessions, normally lasting less than a week, reduced later to two per year. More time was allocated to the Working Groups, in particular to WGB, which initially met three times per year in sessions of two weeks each. Gradually the meeting times were also reduced for the Working Groups. The meeting schedule was at times subject to considerable discussion among the States Signatories, some of which saw the need to limit the meeting time.

Initially the chair of the PrepCom rotated twice a year among the regional groups. After some years it was realized that this did not provide much stability and leadership to the PrepCom and the term was increased to one year. PrepCom chairs are listed in Annex 2. Among the PrepCom chairs is the current 8th Secretary-General of the United Nations, Ban Ki-moon, see Fig. 9.2. The chairs of the two Working Groups initially had no time limit to their tenure and the first two chairpersons served for some nine years. It was then agreed that the Working Group chairs should be appointed for a specific term of two years for WGA and three years for WGB. The chair of the Advisory Group has served since 1997 after several reappointments for three year terms.

Fig. 9.2 Eighth Secretary General of the United Nations, Ban Ki-moon of the Republic of Korea, served as the PrepCom chairperson during the first half of the year 1999

9.1.2 The PrepCom Plenary Body

The plenary session of the PrepCom is the only forum within which formal decisions can be taken. The two Working Groups prepare issues for decision by the PrepCom plenary sessions and most issues are thus in practice settled in the Working Groups. Only the annual budget, the meeting schedule and politically controversial issues have provoked any extended discussions in the plenary sessions of the PrepCom.

To establish a new international, highly technical organization in a political environment is in many ways a challenge. One essential element is to strike a balance in the interaction between States Signatories, the Policy Making Organs and the managerial level of the organization. The relation between States and the PrepCom and its PTS is discussed in the next chapter. The relation between the Policy Making Organs and the PTS has developed over time. During the first years most of the competence and resources were with the Policy Making Organs and a lot of work was done in the two Working Groups to establish plans, technical specifications and administrative and financial rules and regulations. Eventually the PTS grew in strength to become the driving force, which is a natural development. The Policy Making Organs could then focus on their main task of policy making and oversight.

The real influence of the PrepCom plenary body has thus varied over time, partly depending on who was in the chair. Some chairpersons have been very much engaged and showed leadership and were also able to significantly influence the work of the PrepCom and the PTS; others were less engaged and influential. Interest and engagement also vary significantly among the delegations. A few delegations are generally quite interested and knowledgeable and influence most issues. Some delegations follow just one or a few issues of particular national interest, while many other delegations hardly participate actively in any discussion. Positions and statements have normally been formulated by the regional groups or by other groups such as the EU and G77. States coordinating those groups are then active during the tenure of their chairmanships, together with a few other States which have an interest in particular issues. People matter in the PrepCom, too, and sometimes delegates, including those from small delegations, display an impressive insight and engagement, in this way gaining significant influence on a number of issues. What we have witnessed here might mirror the picture in most international organizations and multilateral negotiations.

The mood in the Policy Making Organs was initially most constructive, almost enthusiastic. Many came from tight negotiations in Geneva and now faced the challenge of implementing what they had finally agreed on and making sure that what they fought hard for in the negotiations was not lost in the implementation. All were determined to work constructively together, searching for solutions, not obstacles. Even once the initial optimism had dissipated, the work progressed by and large constructively. Over time the political differences have become more pronounced between States who want to keep on a fast track towards completion of the system and those who want a slower pace. A growing tension has recently developed between the EU and the G77 group. Politically controversial issues unrelated to the PrepCom's task have been introduced with increasing frequency, and such actions have influenced the work and the mood in the Policy Making Organs.

9.1.3 The Working Groups

The two Working Groups chose somewhat different ways of working. WGA had, especially in the beginning, a number of issues to be settled, mainly related to administrative rules and regulations. Friends of the Chair and focal points were appointed to deal with these individual issues as they emerged. WGB chose a more process-oriented way of working, with an internal infrastructure that was stable over time. Some ten task leaders, experts in their respective fields, have been responsible for individual processes, such as the five verification-related programs or the various operational manuals. A list of the experts who have served the Working Groups in this way is provided in Annex 2. These processes remained essentially the same over the years, even if the content changed. A lot of WGB time was devoted to elaborations under the guidance of these task leaders. The WGB cair, the two Friends of the Chair and the WGB secretary focused on keeping these processes on track, through close coordination with delegations, the PTS leadership and the task leaders. This team of task leaders, working closely together, represented an impressive knowledge base and thanks to an ongoing professional dialog with their counterparts in the PTS, they have provided a lot of informal guidance.

The Policy Making Organs are using a program structure to provide program and budget guidance and to exercise its oversight responsibility. As in many other organizations, the PrepCom plenary and its Working Groups tend to devote most effort and interest to planning but less to follow-up. As we found that it was difficult to get WGB engaged in the follow-up of PTS activities, we decided to get external help. In the scientific world peer reviews by outside experts are an essential element in reviewing an organization or a research program. We noted that such external reviews are not a normal tool in the assessments of international organizations, and both States and the PTS were surprised at our idea of holding peer reviews of the technical programs. After initial hesitation they all went along, with increasing interest, and such reviews (CTBT/WGB-14/INF.3 2000, CTBT/WGB-17/INF.3 2001, CTBT/WGB-21/INF.5 2003) were conducted of the three major technical programs covering the International Monitoring System, the International Data Center and On-site Inspection. The reviews and their results, which are discussed later in this chapter, turned out to be most useful and were greatly appreciated by the Policy Making Organs and the PTS. A large number of the recommendations have been implemented and have contributed to making PTS a more cost-effective organization.

9.1.4 Executive Secretary's Two Faces

The Executive Secretary (ES) has two important functions within the PrepCom: that of Executive Secretary of the PrepCom and the head and chief executive officer of the Provisional Technical Secretariat. The central person in a political body like the PrepCom and the leader of a highly technical international organization like the PTS are two quite different tasks, each of which is demanding in its own right. To be the ES of the PrepCom is very much a high- level, international diplomatic task, requiring coordination with States Signatories in Vienna and abroad, supporting the chairs of the PrepCom and the Working Groups and facilitating the smooth functioning of the PrepCom. This is a position well suited to a senior diplomat, and Dr Hoffmann, the first ES, and Mr Tóth, his successor in 2005, are both experienced diplomats (Fig. 9.3). Dr Hoffmann was German ambassador to the CD and played an active role in the negotiation of the CTBT. Mr Tóth was Hungarian ambassador to the CD and chaired the negotiations on the BWC protocol as well as the BWC review process. He was also WGA chairman from the beginning of the PrepCom until his appointment as ES.

To lead a fairly large international organization with a complex technical mission is quite another challenge, requiring managerial experience and skill

Fig. 9.3 Executive
Secretaries of the CTBTO
Preparatory Commission
Wolfgang Hoffmann (*left*)
and Tibor Tóth (*right*)
meeting again at the
Synergies with Science
conference in 2006

as well as a general understanding of technical and operational issues. We shall return to a discussion of this aspect of the ES function later in this chapter.

9.1.5 How did it Work Out?

From an organizational and managerial point of view, how have the Policy Making Organs of the PrepCom functioned? As we were heavily engaged in that work ever since the start of the PrepCom, we might not be objective judges, but we would like to share some of our experience.

The structure of the PrepCom, with two Working Groups dealing with defined tasks, was a good one from the beginning. Initially a number of issues of a technical and administrative nature needed to be dealt with and this was most efficiently done in different forums with participation of the proper expertise. The number of such specific issues has decreased over time. It also proved useful to handle some issues, such as budget and organizational issues, in joint meetings of the two Working Groups. The need for two separate Working Groups is thus gradually dissipating. After entry into force there will also be only one body, the Executive Council, which will support the Conference of the States Parties.

As the establishment of the verification regime is the main task for the PrepCom, it has, through WGB, been heavily engaged in a number of technical issues. The internal structure of WGB, with a team of task leaders having responsibilities for specific processes, has proved to work well, in our judgement. The initially very high level of ambition was certainly most useful and necessary to get activities going. WGB also saw the need to maintain, over a longer period of time, considerable engagement in the coordination and planning of the technical activities in the PTS. WGB's involvement in some issues could, in hindsight, have been reduced more rapidly as time passed.

It is not obvious what constitutes a good balance in the division of work between the Policy Making Organs and the PTS, and this balance may also change over time. The Policy Making Organs should strive to provide guidance and oversight at the right level of detail. Goals should be set that are easy to follow up and to measure progress against. The PTS must report its progress related to those goals and milestones. The organizational review, discussed further below, notes that "*Clearly identified budgetary targets, related to goals and milestones and with enough details to define responsibility, should be part of the budget process. The targets would form the basis for reporting on performance to the Policy Making Organs*". We agree that the Policy Making Organs must spend more efforts on the follow-up process and give clear feedback. The PTS management level should also be held more directly responsible for the results achieved. To establish and operate an organization with a complex technical task in a political environment is a challenge shared by the States and the PrepCom. It requires trust and cooperation between the States, the Policy Making Organs and the PTS, based on respect for their respective roles.

All in all, and given that the PrepCom phase has lasted far longer than originally anticipated, we feel that the Policy Making Organs have functioned well and fulfilled their duties. We also feel that there has been a constructive relation and dialog between the Policy Making Organs and the PTS leadership, both at the Executive Secretary and director level. The informal interactions among experts from WGB and the PTS on a number of technical issues have also been most useful.

9.2 Provisional Technical Secretariat (PTS)

The resolution establishing the CTBTO Preparatory Commission says very little about the PTS and how it should operate. The resolution just states that "*The Commission shall establish a provisional Technical Secretariat to assist the*

Commission in its activity and to exercise such functions as the Commission may determine, and appoint the necessary staff in accordance with the principles established for the staff of the Technical Secretariat pursuant to article I,I paragraph 50, of the Treaty. Only nationals of States Signatories shall be appointed to the provisional Technical Secretariat". Based on this general directive, and with no specific guidance on its size or how it should be organized, the PrepCom and the PTS have developed the frame needed for the PTS and its extensive activities in Vienna and around the world.

9.2.1 The First Step

Following a lot of informal consultation before and during the first PrepCom session, which started in New York in November 1996 and ended after a break in Geneva in March 1997, Dr Wolfgang Hoffmann was appointed Executive Secretary of the CTBTO Preparatory Commission and chief executive officer of the PTS.

The first PrepCom session also agreed on a PTS, organized in five divisions with an anticipated staff of around 300. The discussions leading to this organization illustrated that we were creating a technical organization in a political environment. The number of divisions were decided more on political than managerial grounds, to give each of the key players a piece of the cake. A common understanding was developed that the directorship of a certain division should go to a particular group of States: Administration to North America and Western Europe, Legal and External Relations to South East Asia, the Pacific and the Far East, IMS to Latin America and the Caribbean, IDC to Africa and OSI to Eastern Europe. Until the end of 2007 the directorships have been distributed among States as follows: Administration: three US directors; Legal and External Relations: Japan and China; IMS: Mexico and Costa Rica; IDC: Egypt and Burkina Faso and OSI: two Russian directors. The Middle East and South Asia group, paralyzed by internal difficulties and unable even to meet, was not given the responsibility for any division. It is not only at the director level that political aspects play a great role in recruiting staff members.

From the beginning some were concerned that it might not be possible to find a single individual with the experience required to handle the diplomatic functions of the Executive Secretary of the PrepCom and those of managing a technical organization. Proposals were therefore put forward during the consultations leading to the first session of the PrepCom, in 1996, to create a senior position in the ES office to support the ES on technical issues. It was proposed that the person in that position should be responsible, under the ES, for the technical verification work and thus guide and coordinate the technical divisions. Such a managerial structure would have made it possible to combine diplomatic, managerial and technical experience at the management level of the PTS. This idea was rejected by many States Signatories and by the incoming ES.

9.2.2 A "Family" Turning into an Organization

Many of us who had been involved in the negotiation of the CTBT for many years in Geneva moved to Vienna in early 1997. Dr Wolfgang Hoffmann, the first Executive Secretary, carried the banner for the nine staff members who were the first to join the PTS. Many experts from Geneva became national delegates to the PrepCom, in particular to WGB. The Policy Making Organs quickly got up to speed and a lot of the work during the first years was also carried out in the two Working Groups. The initial PrepCom phase, when we all expected an early entry into force, was filled with enthusiasm and excitement. We all felt the challenge: how would we manage to get everything in place in time for entry into force? The Working Groups and the PTS were working closely and informally together to get the organization going. Our common experience was that we were a family facing a common challenge.

The US decision in October 1999 not to ratify the Treaty was a cold shower that lowered the mood and the enthusiasm. It was interesting to note, though, that no country took any concrete action to demonstrate its discontent. Business went on as usual but the sense of excitement and urgency was gone. We all knew that we were in for a long journey.

Eventually PTS had grown to a size and a strength of a full-blown organization. This was also the time when the informal and "family" way of working did not work any longer. There was a need for a management system, for coordinated planning and for managerial skills at all levels of the organization. From its beginning the PTS managed to attract a number of world-class scientists, creating a good scientific basis for its work. Few people with prior senior managerial experience were hired, though. This lack of managerial experience throughout the organization gradually became evident.

9.2.3 Evaluations Push for Better Coordination

The lack of coordination between different technical activities became increasingly evident and was clearly observed and noted in the reviews of the individual programs and of the organization as a serious impediment to efficient, cost-effective work. Even the first external review, the one on the IDC program in 2000, noted that "*consideration be given to a new senior management appointment to take responsibility for the operational management, integration and strategic planning of the verification function as a whole*". The organizational review notes, five years later, that "*Consideration should be given to enhancing the coordinating role of the Office of the Executive Secretary with the purpose of more effectively managing the PTS as a whole*".

These difficulties in coordination and planning were noted in all the reviews and expressed in somewhat different ways. The review of the organizational structure (CTBT/PC-24/INF.9 2005) "*identified needs common to all areas of the*

PTS for enhancement of managerial processes. It underlines the importance of strong, coherent approaches to goal setting, integrated programme and budget planning and implementation and human resource management, all supported by well defined evaluation measures. Clearly identified budgetary targets, related to goals and milestones and with enough detail to define responsibility should be part of the budget process". The evaluation of the OSI program noted similarly that *"Planning should be strengthened, flexible and subject to regular reviews and it should be focused on a specific goal which should be a fully comprehensive field exercise"*. The evaluation of the IMS program stated that *"An unfortunate lack of coordination and cooperation was noted between the four technology sections in the IMS and between the IMS and the IDC Divisions"*. This significantly affected the ability of PTS to make credible plans that could be fully implemented. For several years the PTS, for a number of reasons, fell short of building and certifying the number of stations that was planned in its annual work plans. The lack of proper managerial processes made it difficult for the PTS to conduct its work in a project-oriented, well-coordinated way. Each division was very much operating on its own, and this "stove-piping" was of growing concern to the Policy Making Organs. In this area of decision making and method of work, the organizational review *"recommends policies and coordinating procedures to manage the functions of a predominantly operational organization and allow for greater integration of activities. Supported through the Office of the Executive Secretary, these could include further development of integrated strategic planning, coordinated, inclusive mechanisms for decision making, an integrated oversight and reporting system and more extensive use of performance measurement and project management techniques"*.

The organizational review also contained a number of comments and recommendations on how to adjust the PTS organizational structure to better meet future needs. These recommendations formed the basis for the new organization discussed below. In that regard the review team also noted the clear link between the Policy Making Organs and the PTS and that *"the PTS structure can not be considered in isolation from the working arrangements of Policy Making Organs"*. The IMS review team also made a similar observation: *"The Programme structure and the PTS organizational structure also mirror each other, reinforcing this tendency for PTS sections and divisions to operate as quasi-independent units rather than as components of an integrated whole"*. The PTS reorganization, discussed later in this chapter, aligned the technical divisions and the verification programs even more than before. We very much share the observation that it is unfortunate that the program structure and the PTS organization mirror each other and thus counteract internal integration. Efforts have failed over the years to create a structure in which the programs cut across the organization rather than being aligned with the divisions. Such a structure, with each program involving many parts of the organization, promotes close internal cooperation and the creation of clearly defined projects.

An important experience, documented in several reviews, is thus that international organizations, in the same way as national ones, depend critically on an orderly planning and decision structure and working practice. They also need staff members with good managerial skills and experience at all levels in the organization. This acute need for managerial skills and experience is most likely to exist in any organization. The need might be even greater in an international organization, with a more heterogeneous staff composition than in a corresponding national organization. We all understand and appreciate that political considerations are essential elements when creating an international organization that will operate in a political environment, but we are convinced that managerial considerations are equally important. To achieve a well functioning organization requires careful consideration by States Signatories when establishing the organizational structure and the rules and regulations that guide its work. It also requires great care by States Signatories when promoting people to staff positions, and by the leadership of the organization that does the recruiting.

9.2.4 Rules and Regulations

The work of the PTS is guided by a large number of rules and regulations, covering some 200 pages in all (CTBTO Preparatory Commission 2005), which were established by the PrepCom at an early stage. These rules are supplemented by over 50 administrative directives issued by the ES. To a great extent these regulatory documents mirror those of the UN and other international organizations. Their prime purpose is to promote efficient, fair management and operation of the PTS and provide for proper oversight by the Policy Making Organs.

Reflecting the environment in which they have been created, rules and regulations sometimes tend to be overly formalistic and legalistic rather than having a managerial perspective. The IMS review team noted that representatives of States Signatories and the PTS staff had highlighted the procurement function "*as being too slow, lacking in flexibility and too legalistic. The team tends to agree with these observations*".

It has been said, and wisely so, that you cannot run an organization by committee. PTS, and surely most other similar international bodies, is run according to rules and regulations created by international committees. In the case of PTS the body is WGA, where representatives from some 100 States with very different economic systems and traditions to manage organizations have tried to find a lowest common denominator. Moreover, few members of such a body are likely to have any personal experience of managing a fair sized organization. So what is the result? Some would no doubt conclude that the outcome may not be the best possible basis for an efficient management of the PTS. Others may agree with James Surowiecki, who in his "The Wisdom of Crowds" (Surowiecki 2005), argues that a large group of people, even without expert knowledge in the actual area, can provide better solutions than a small team of experts. The sad thing is we do not really know how good or bad the rules or regulations are, and what the consequences are when they have been implemented, as this has never been evaluated. The consequences of two of the rules or administrative directives have become clear, as they have created considerable difficulties: one is a regulation on handling the payment of arrears; the other an administrative directive on the seven-year service limit. We discuss these two issues in the following.

The regulations could sometimes have serious, and we assume unintended consequences. The regulation for handling payments by States that are in arrears is a good illustration of what can happen. When a State has not paid its assessed contribution for one or more years, or paid only part of the amount, a burden of debt is created. This accumulated burden of debt could go back many years: for those who have never paid, it by now extends over more than ten years. The statements of accounts for all but the last year have been approved long ago and the books have been closed. When a State having a debt starts paying its contribution again, it has to cover that debt, and if the books are closed the money cannot be used by the PTS but has to go back as a surplus to all States Signatories that have paid. In this way the PTS is prevented from using for programme implementation any money that is paid after the year in which it was due, as illustrated by the USA situation: as of September 2007 the USA owed the PTS approximately $25 million related to assessed contribution for 2007 which the USA had not paid in full. Under the current financial rules and regulations, payments from the USA received in 2008 would be credited to the 2007 arrears. In effect, it would be returned to States as part of the 2007 surplus. It would thus not be available to the PTS for implementing its 2008 programme. Realizing this consequence, the PrepCom at its session in November 2007 decided to *"extend the 2007 annual obligating authority from the General Fund until December 31, 2008"*. The result of this and other related decisions was that, in fact, all of the almost $24 million that the USA paid in February 2008 can be effectively used for program implementation. This need to take extraordinary measures is to us a good example of a rule with very strange and, we assume, unintended effects.

We acknowledge that the rules and regulations for the PTS had to be established in great haste at the beginning of the PrepCom and that nobody at that time expected that they would be used for a long period of time. It might also have been difficult to recognize at that time the managerial and practical consequences of the rules and regulations agreed. It thus seems essential that they be reviewed to reflect experiences gained within the organization and general developments in our societies. PrepCom has conducted external reviews of the three individual verification major programs and of the organization itself. No such external review has so far been undertaken to address the rules and regulations and the way they are functioning in practice. Hopefully the PrepCom period will soon come to an end, but it would still be most useful to conduct a review of the rules and regulations to prepare for a cost-efficient and smooth functioning of the CTBTO and its Technical Secretariat after entry into force. Given that many of those rules and regulations are very similar in many organization there might be an advantage of conducting such a review on a broader scale including several similar international organizations.

9.2.5 Staffing and the Seven-Year Service Limit

Despite the political constraints on the recruitment process, the PTS attracted world-class scientists and other experts from around the world. Many were thrilled by the challenge of participating in the build-up of a unique verification system and working in an international environment. The organization was growing gradually, and by 2001 the PTS was essentially fully staffed with personnel of about 270 (see Fig. 6.12). In 2002, the PrepCom plenary body instructed the PTS not to create any new positions beyond the existing 286 in the budget, and new needs have been filled since then by rearranging the existing positions.

An international organization is a thrilling, creative environment, but not necessarily a very simple one to work in as people come in with different backgrounds, experiences, traditions and expectations. We have already noted that the PTS established a solid scientific basis for its work by engaging a number of world-class scientists, but also that managerial experience was in much shorter supply throughout the organization. As political considerations played an important role in appointing persons, especially to managerial positions at different levels, it could happen that more experienced and qualified persons became subordinate to less experienced chiefs.

Any organization repeatedly states that its personnel are its greatest asset and most managers also sincerely believe that. How has it worked in practice at the PTS, then? In 2001 the IMS review found "*an unacceptable tension between many of the IMS staff and the Personnel Section*" and the review noted that "*the Personnel Section is perceived to play a line management role in the approval of several personnel actions e.g. hiring of staff*". The review team "*believes that these are properly the responsibility of the PTS line management, with the Personnel Section acting in an advisory capacity*". In any organization it can quite easily happen that the administration unit takes on the function of managing the organization rather than serving it. The human resource area was a clear example in the PTS, but not the only one. Other functions, too, such as legal and procurement, took on responsibilities that in our view belonged to the line management. One of the reasons for this was no doubt the lack of a good managerial structure and practice. The organizational review team "*underlines the continuing urgent need for the Executive Secretary and the PTS to complete a comprehensive human resource plan with more precise job descriptions so as to manage the non-career policy and to meet the changing needs of the PTS*". The team stressed that "*human resource planning is a critically important part of any organization. And, given the seven years service limit, there is a continuing urgent need to develop a comprehensive human resource plan*".

We turn now to the seven-year service limit and its application, which over the years has sapped so much energy from the PTS. It was decided from the beginning by States Signatories and stated in the "Regulations and Rules" (CTBTO Preparatory Commission 2005) that "*Staff shall be granted fixed-term appointments under such terms and conditions, consistent with the present Regulations, as the Executive Secretary may prescribe*". It also states that "*In granting fixed-term appointments, the Executive Secretary shall bear in mind the non-career nature of the Commission*". States thus established that the PTS should be a non-career organization, meaning that you should spend only a limited part of your professional life there and that you should not expect to be promoted within the organization. It also decided that the contracts should be time limited but did not regulate the length of each contract or the overall service limit for the professional staff. There is no service limit for the general service staff, and here too the PTS attracted most competent and experienced support personnel. The Executive Secretary, in an Administrative Directive issued in 1999, set the service limit for the professional staff at seven years. He also decided that the initial contract should be for three years, which could be extended twice by two years. In reality all contracts were, almost automatically, prolonged to the maximum length of seven years. At a number of sessions of the PrepCom States have expressed the importance of observing the provisions of the seven-year service limit. The system for personnel performance assessment that was put in place has not been fully utilized and personnel assessment has not been documented in a way that satisfied the scrutiny of the International Labor Organization (ILO) tribunal. No plan was created to gradually renew the staff, and eventually a large number of staff members had reached the seven-year service limit, creating a great problem when a large percentage of the professional staff turned over in a short period.

This seven-year service limit and the way it has been applied was cause for great attention and concern among the professional PTS staff. The application of this limit was challenged before the ILO Tribunal, which ruled in favor of the staff members, as the seven-year service limit spelled out in the Executive Secretarýs Administrative Directive of 1999 was not formally stipulated in staff contracts. As a result, the Executive Secretary granted staff members another extension of two more years. Now they could serve nine years, but then what? To cope with the high turnover after nine years, an opportunity has also been opened to staff members to compete for their positions with outside candidates. If you win that competition you will be granted up to another two years, now eleven years in all. And what will happen then?

This Administrative Directive and the way it has been applied has created a lot of confusion and frustration among the PTS staff and in different ways has been costly for the PTS. A number of cases have ended up at the ILO Tribunal, causing PTS to lose some of its good initial reputation and resulting in fewer job applications from experienced people. This clearly illustrates the importance of having employment conditions that are clearly understood and accepted by all. It may be even more important that they are applied in a manner that is considered fair, evenhanded and transparent by the staff. An international organization that is not perceived as handling its staff respectfully and fairly will soon find itself lacking the needed expertise.

Is this seven-year service limit really needed, or even desirable? Before anyone reached the seven-year service limit, annual staff turnover was about 11% for several years. A seven-year limit corresponds to 14%. A healthy and desired renewal of staff was thus already occurring, without that Aministrative Directive. We agree that persons in decision making or otherwise influential positions should have time limited terms to ensure that the organization is revitalized from time to time. This will also minimize the risk of the build-up of "local empires". In our view the time limited contracts of three and two years provide the necessary tool, if they are fully used together with a staff appraisal system. On the other hand, PTS has a number of specialists, such as the data analysts, who need many years, maybe as many as ten, to train fully. We see no reason why those experts should have a limited service time, especially as there are few job opportunities outside the PTS. For other professional staff members, too, we think PTS should work on an individual basis together with their home institution, often a university or a research organization, to ascertain that the time a person spends at PTS is not only a useful contribution to the PTS but also a rewarding element in their personal career.

9.3 Changing the Guard

In August 2005 Dr Wolfgang Hoffmann left after having served as ES for more than eight years and Ambassador Tibor Tóth arrived. Mr. Tóth was replaced as chairman of WGA by Ambassador Abdul Bin Rimdap from Nigeria, a former chairperson of the PrepCom. In March 2006 Dr Ola Dahlman left the chair of WGB, which he had held since the beginning of PrepCom, to be succeeded by Dr Hein Haak, who for nine years served as a Friend of the WGB chair (Fig. 9.4).

Fig. 9.4 Group photo of the task leaders and chairpersons of Working Group B, taken at the Vienna International Centre on 3 March 2006 on the occasion of the handover of the chairmanship from Ola Dahlman to Hein Haak. *Left* to *right*: Kazuto Suda, Hans Frese, Bernard Massinon, Vitaliy Shchukin, Arne Bell (Secretary of WGB), David McCormack, Mohsen Ghafory-Ashtiany, Hein Haak, Ola Dahlman, Svein Mykkeltveit, Jay Zucca, Malcolm Coxhead and Frode Ringdal

The changing of the guard was most pronounced throughout the organization, both in the Working Groups and in the PTS. The task leaders, comprising the backbone of WGB, have gradually been changing over the years and few who served in 1997 were still there in 2007. The turnover has been even greater in the PTS, where all the directors are new in 2007, compared to 1997, and a great majority of the staff has turned over. This turnover was intended as a result of the seven-year service limit, as we have discussed.

9.3.1 Reorganizing the PTS

Not only has the staff changed, so too has the way the PTS is organized. The PrepCom decided in 2004 to review the organizational structure of the PTS, which had been in existence for seven years, during which time its activities had gradually shifted from a build-up phase to one of testing and evaluation. The impending replacement of most of the senior management, including the ES, and an expected large turnover of personnel due to the seven-year service limit also offered an opportunity to make organizational changes. The review (CTBT/PC-24/INF.9 2005) was conducted by an international team of senior experts from academia, industry and the diplomatic field under the able leadership of Richard Starr, former Australian ambassador and one of the key CTBT negotiators in Geneva, and Ambassador Abdul Bin Rimdap of Nigeria, who was later elected chairperson of WGA.

The review recommended "*the redefinition of the functions of the IMS and IDC Divisions in order to better align processes with the shifting functions of the PTS*". The review team recommended that the IDC and IMS divisions be replaced by an operations division, and an engineering, development and logistics support division, respectively. The operations division should "*be responsible for operation of the infrastructure and processes necessary for the continuous collection, transmission, analysis and dissemination of data and data products, and related testing and evaluation*". The division for engineering, development and logistics support "*should bring together the technical functions that support PTS operations*". The new organization, based essentially on the recommendations in the review, was created in 2007 and is shown in Fig. 9.5. The staff of the operations division, still named the IDC division, is about 100 and that of the engineering division, still called the IMS division, is about 50.

We believe this reorganization will prove useful for the test and evaluation phase of the verification system, as it is gradually coming into provisional operation. We further see this as an important step towards an organization that can be smoothly transformed into the Technical Secretariat once the Treaty enters into force. It will then be essential to streamline the organization, giving the IDC division the task of providing services to States Signatories. The division could, in shaping its processes, benefit from experiences of service providers in other areas. An important element for the organization is to

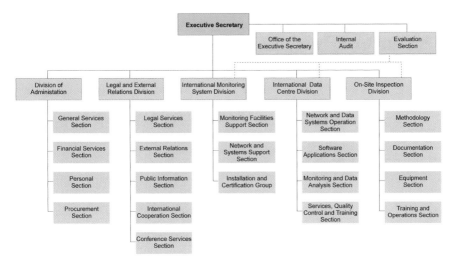

Fig. 9.5 Diagram showing the PTS structure after completion in 2007 of the reorganization

make full use of new knowledge, technologies and analysis methods and procedures developed in the S&T world. This is essential to keep the organization viable, to continue to attract good people and to constantly develop the cost-efficiency of the organization and the quality of its products. This "development" part of the IMS division is thus a most essential element that has to be developed considerably in close contact with the scientific world and high-tech industry. In the new organizational structure, the IDC division has been given the responsibility of analysis software development. In our view it is not a good idea to give an operational division the responsibility for development work. All such development work should, in our view, be the responsibility of the IMS division, which would also be responsible for cooperation with the scientific community. We discuss the cooperation issue further in the next chapter.

Working Group B has also modified its method of work to create a clearer distinction between policy guidance and oversight. The intent is that this will on the one hand facilitate more focused considerations of policy guidance to be given to the PTS in terms that are possible to implement and monitor. On the other hand it will also give a clear focus on oversight that will hopefully allow WGB and the PrepCom to spend more time on the follow-up process.

Today's PrepCom and PTS are quite different from those of a decade ago. When the PrepCom and the PTS came to Vienna almost all of us who were engaged in their establishment came from the negotiations and had a long-term background in verification work. Today, very few have personal experience of the negotiations or research work on verification. Today's PTS is an organization that is less influenced by history but hopefully has a fresh approach to how to operate a complex technical organization in support of the implementation of the CTBT.

Chapter 10
The CTBTO Preparatory Commission and the World

The CTBTO Preparatory Commission has close relations with many global actors. States Signatories are the organization's masters, providing the funds needed as well as getting back investments in the monitoring stations they host. States provide personnel to the PTS, which has an extensive program for training national experts. States are also the customers for the data and products produced by the PTS during the testing and provisional operation of the emerging global monitoring network.

The PrepCom is part of the close cooperation on logistical matters among the organizations at the Vienna International Centre (VIC). On other matters the cooperation with the Vienna-based organizations is, for different reasons, less extensive than many had anticipated during the negotiations. The PrepCom has a number of cooperative agreements with international organizations. Formally the most important is that with the UN, the Secretary-General of which is the depositary of the CTBT, while the most important from a practical point of view is with the World Meteorological Organization.

The PrepCom is a high-tech organization that needs to develop in synergy with science and modern technology. The system that is being implemented is designed on the basis of scientific knowledge available at the time of the CTBT negotiations. To keep the system and the organization viable there is a need for a continuous influx and application of new knowledge and modern technology. To develop and sustain close ties with the scientific community is not a luxury that the PrepCom can afford or not, as it pleases: it is a truly strategic relation.

Like any other international treaty and organization, the CTBT and the PrepCom are hostage to the actual political situation. This is unavoidable, but it might be interesting to consider how to increase the resilience of international agreements and activities. One way might be to reduce the consequences of actions by individual States by evening out the responsibilities and the commitments.

O. Dahlman et al., *Nuclear Test Ban*, DOI 10.1007/978-1-4020-6885-0_10,
© Springer Science+Business Media B.V. 2009

10.1 States Signatories, the CTBTO Preparatory Commission and the PTS

We have earlier noted that, in many ways, it is a challenge to establish a new international, highly technical organization in a political environment. In this chapter we discuss the roles and tasks of the key actors and efforts over the years to find a balance in the interaction between States Signatories, the CTBTO Preparatory Commission and its Provisional Technical Secretariat, the PTS.

The States Signatories have signed the Treaty and they are responsible for the Treaty, its implementation and entry into force. The PrepCom represents the collective effort by the States to achieve these goals. As part of the PrepCom, PTS is a service organization for States Signatories and the PrepCom, assisting in the implementation of the CTBT. This is the bottom line, but still there are a number of different facets to the relation. It takes a careful, humble attitude on the part of both States and PrepCom and its PTS to handle such a multitude of roles and relations in potentially conflicting situations. A large number of practical issues of a legal, economic and technical nature have come up between individual States and the PTS during the build-up of the verification regime. Today's realpolitik has no doubt also influenced the PrepCom's work. Even if problems have surfaced at times, our impression is that the representatives of States Signatories and the PTS leadership have over the years handled their relations professionally and fairly.

10.1.1 States are the Masters of Implementation

States Signatories have the overall responsibility for the implementation of the CTBT and they are also responsible for promoting its entry into force. They may use different means, as they see fit, to persuade other States to sign and ratify the Treaty. Article XIV of the Treaty states that if the Treaty has not entered into force three years after it was opened for signature, a conference should be held to consider how to accelerate the ratification process to facilitate the Treaty's entry into force. A number of such conferences have been held, in many cases with high-level attendance. The concrete results of these conferences have been limited at best, and it is unlikely that such general political manifestations alone will provide a breakthrough in bringing the Treaty into force.

States are the masters of the PrepCom and its PTS. In the Policy Making Organs of the PrepCom they take the decisions needed to guide the implementation of the CTBT and the work by the PTS. States decide annually on a work program and budget for the PTS. They also decide on the overall size of the staff and they appoint the Executive Secretary and the directors. They can interfere in any issue as they see fit. By and large, States are quite restrained, limiting their engagement to defining what should be done and leaving it to the PTS to decide how to do it. Sometimes, however, they do get involved in the details of a particular issue in ways that are perceived by the PTS as micromanagement. The PTS complaints about micromanagement were at times justified, while on other occasions such complaints reflected an overly sensitive attitude.

10.1.2 States are Providing – and Getting – Money

States are the financiers of the PrepCom including the PTS. States have by and large proved ready to grant the PTS the funding it has requested, which for a number of years turned out to be more than the PTS managed to spend. After a gradual increase, the PTS annual budget has stabilized at just above $100 million (Fig. 10.1). Including 2007, $900.8 million in all has been allocated in the annual budgets. The budget is allocated to six major programs and divided into a general fund, and a capital investment fund, for which money can be carried over from one year to the next. Table 10.1 shows the annual budgets and expenditures and their allocations to verification and non-verification programs. Some 80% of the money has been spent on the technical programs, in support of the establishment of the global monitoring system and the OSI regime, and some 20% has gone into administration and to support the sessions of the Policy Making Organs.

A large part of the PTS budget goes to investments in States hosting IMS facilities. The amount of money that will be spent on the total of 337 such facilities is currently estimated at $328 million in all (CTBT/PTS/INF.872 2007). This is money spent on investments in the States Signatories. Most of the investment money will be used for infrastructure, and local construction and installation work. Once a station is installed and certified, the PTS concludes a

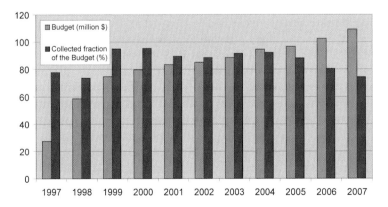

Fig. 10.1 Annual budgets and percentage of the budget collected in the period 1997–2007. The vertical axis is both in million US$ and in percentages

Table 10.1 PTS annual budgets and expenditure in millions of US$

Financial year	Budget			Expenditure		
	Verification	Non-verification	Total	Verification	Non-verification	Total
1997	15.80	11.70	27.50	4.00	9.80	13.80
1998	42.10	16.30	58.40	20.10	12.70	32.80
1999	59.70	15.00	74.70	56.40	13.50	69.90
2000	64.00	16.00	79.90	59.90	13.80	73.70
2001	69.20	14.30	83.50	79.70	13.60	93.30
2002	70.80	14.30	85.10	58.20	13.50	71.70
2003	73.70	14.90	88.60	71.00	15.70	86.70
2004	76.20	18.40	94.50	65.50	18.20	83.70
2005	77.10	19.50	96.70	82.10	20.40	102.50
2006	80.90	21.60	102.50	74.40	20.90	95.30
Sub-total	629.40	162.10	791.50	571.30	152.20	723.40
2007	86.70	22.60	109.30	–	–	–
Total	716.10	184.60	900.80	–	–	–

Notes: From 2005, a split currency system (Euro/US$) was introduced and amounts in this table reflect US$ equivalent. Differences in totals may reflect rounding differences.

contract with a local operator to maintain and run the station. This operator is normally a scientific institution that in this way is supported in building up and maintaining a sustainable activity. The operator can also use the data from IMS stations for its own purposes. The total amount of money to be allocated to operation and maintenance once all stations are installed is estimated at around $45 million per annum, including recapitalization costs. Most of those monies are spent in States Signatories. States also host a number of workshops and training activities, organized and paid for by the PTS.

A few years after it was established, the PTS eventually collected more than 95% of the agreed annual budget, a collection rate considered to be fairly high compared to other similar organizations. Over time it decreased and for 2006 the collection rate was as low as 80.5% (Fig. 10.1). In 2007 the collection rate was even lower, 74.5%, due to the limited US payment applicable to this year.

Before 2007, our experience was that PTS never lacked money for its activities, year on year. The problem facing the PTS is that not all the money is available at the beginning of the fiscal year: payments by States arrive more or less at random during the year. As the PTS has been careful not to spend money that was not in the bank, this has created a severe problem as contingency margins were imposed that spoiled the planning and left a fair amount of money unused at the end of the year. Money that has not been used when the annual accounts are settled, or payments made by States after that date, have to be returned to States in proportion to the amount they pay.

States provide money according to the UN scale of assessments, slightly amended as not all UN members are signatories to the CTBT. The USA has the highest assessed contribution of 22.3%. Japan contributes 19.7% and the joint contribution from the EU States comprises 37.1%. The principal share of the money is thus provided by a fairly small number of States. 44 States have the minimum assessed contribution of 0.001% of the total budget or about $1000 per year, given the present PTS budget of just over $100 million. A large number of States, most of them with minimum assessed contribution, do not pay their contributions. During the ten years since the PTS was created, 42 States have not paid any contributions; 20 of these have the minimum assessed contribution.

Table 10.2 gives the accumulated assessed and paid contributions for the period 1997–2007. This illustrates that the North America and Western Europe group has had an assessed contribution of 63% in all, which it has paid with the exception of the money withheld by the USA: about $24 million (end 2007). The South-East Asia, the Pacific and the Far East group had an assessed contribution of 27% in all, which it has paid in full. The Eastern Europe group has also fully paid its assessed contribution of 3%. The remaining three groups have not paid in full and the Latin America and the Caribbean group has not even paid half its assessed contribution. The accumulated non-payments of these three groups has now reached $27.3 million, which is a substantial amount, corresponding to a quarter of a PTS budget for one year. The lion's share of this comes from four States; Argentina, Brazil, Colombia and Iran with an accumulated debt of $21.5 million.

Since 2002 the USA has substantially limited its payment and has refused to pay for PTS activities related to on-site inspections preparation and international outreach to promote entry into force.

As we have seen, a large portion of the PTS budget goes back to States Signatories in the form of investments in IMS stations and in contracts to operate the stations. This figure stood at $249.8 million at the end of August

Table 10.2 Returned investments for the six regional groups as a function of assessed and paid contributions 1997–2007 in millions of US$

Regional group	Assessed contrib.	Paid contrib.	CIF*	PCA*	Tot. invest.	Return. investm.
Africa	7.7	6.7	16.3	1.5	17.8	266%
Eastern Europe	24.7	24.5	22.6	0.7	23.3	95%
Latin America and the Caribbean	44.4	22.2	31.6	3.4	35.0	158%
Middle East and South Asia	11.3	9.3	17.6	1.5	19.1	205%
North America and Western Europe	582.6	556.3	80.2	18.6	98.9	18%
S-E Asia, the Pacific and the Far East	228.4	227.7	45.0	10.7	55.7	24%
Total	**899.1**	**846.7**	**213.3**	**36.4**	**249.8**	–

* CIF stands for Capital Investment Fund for establishment of IMS stations, PCA stands for Post-Certification Activities for provisional operation and maintenance of certified IMS stations.

2007. Table 10.2 shows the return money in relation to assessed and paid contributions for the six regional groups. Since IMS stations are distributed fairly evenly around the globe, so are the investments in monitoring stations, which means that in this way some regions will get back more money than they contribute.

In addition to money for building and operating stations, States gain access to modern technology and training opportunities. This net transfer of money and technology is most pronounced for developing States having the minimum assessed contribution. The CTBT does not provide for a development support program, as the IAEA does for NPT, but nevertheless large resources are spent in developing countries as part of the establishment and operation of the IMS. Such resources could help build and sustain activities in several scientific areas.

10.1.3 States Provide the Personnel to the PTS

As we discussed in Chapter 9, the PTS initially attracted a number of world-class scientists coming from universities and research organizations around the world. Many came on their own initiative, others were promoted by their national authorities. As of 31 December 2007, PTS had a staff of 253 coming from 71 countries. Looking at the geographical distribution of the present staff, we again compare the regional groups. In Table 10.3 we can see a big difference between the regional groups, which may arise for different reasons, such as availability and interest of persons suitable for recruitment, efforts by States to promote their people, and efforts by the PrepCom and the PTS to recruit on as wide a geographical basis as possible. The North America and Western Europe group provides about half the staff, while very few come from the Middle East

Table 10.3 Staff members and national experts trained for the six regional groups in relation to assessed and paid contributions 1997–2007 in millions of US$

Regional group	Assessed contribution	Paid contribution	Staff recruited*	National experts trained
Africa	7.7	6.7	24	571
Eastern Europe	24.7	24.5	31	496
Latin America and Caribbean	44.4	22.2	16	378
Middle East and South Asia	11.3	9.3	9	271
North America and Western Europe	582.6	556.3	126	797
S-E Asia, the Pacific and the Far East	228.4	227.7	31	557
Total	**899.1**	**846.7**	**237**	**3070**

* Staff recruited between Jan. 2000 and Aug. 2007

and South Asia. Some States argue that there should be a reasonable balance between the number of staff members and the contribution paid to the organization. Looking at the staff distribution in that perspective shows a large variation among the regional groups. On the other hand, it is essential to engage people from around the world in the work of the PrepCom and the PTS. Over the years the PrepCom and the PTS leadership have promoted the recruitment and hiring of staff from developing countries, even those that are not paying their assessed contributions. Hopefully all States will see the value of fully participating in the PrepCom and pay their assessed contributions in full and on time.

The resolution establishing the PrepCom states that the recruitment of the staff during the PrepCom phase should follow the principles laid down in Article II of the Treaty; "*The paramount consideration . . . shall be the necessity of securing the highest standards of professional expertise, experience, efficiency, competence and integrity*". "*Due regard shall be paid to the importance of recruiting the staff on as wide a geographical basis as possible*". In Staff regulation 4.2 of the PrepCom it further says; "*and to the contributions of Member States, as appropriate*". There are frequent arguments among States and between States and the PTS on what constitutes "*as wide a geographical basis as possible*" and what is meant by "*contributions of Member States, as appropriate*". Are these ambitions in fact contrary to each other? If so, which one should have priority? We have seen that a large number of States, especially in Latin America and Africa, have not paid any assessed contributions for ten years. In a statement to the PrepCom session in June 2007, the European Union expressed the view that chairpersons in the Policy Making Organs should come from States that have ratified the Treaty and that do not have long-standing arrears. It further called on the Executive Secretary to take a restrictive line in employing staff from countries with suspended voting rights.

We realize that there is a need for both carrots and sticks to make each State Signatory live up to its fundamental obligations in relation to the Treaty, including paying its assessed contribution on time and in full. Arguments have been advanced that the chairpersons of the Policy Making Organs and the Executive Secretary should come from States Signatories that fulfil all their obligations and that the same reasoning may also apply to the senior PTS staff, the division directors, who are appointed by the PrepCom plenary body. We can see the rationale underlying these arguments. On the other hand, it may prove fruitful not to be too dogmatic, as it is essential to maintain a global engagement in the activities of the PrepCom Policy Making Organs and at high levels in the PTS. For other PTS staff, many of them applying for positions in the PTS on their own initiative rather than being promoted by a State, we believe it is essential not to exclude anyone for reasons of non-payment by their countries of origin. On the contrary, it is essential to create a global knowledge base to support the Treaty once it enters into force.

Given these different arguments, is the present balance a good one? If not, what should it be? There may be as many answers to this question as there are

States or groups of States, each promoting their own perspective. This is one of the important challenges for States and the PTS leadership to handle constructively. It is an issue that should be equally important and difficult for all international organizations.

As of 31 December 2007, the representation of women in the professional category at the PTS stands at 32.9%. This number can be compared to those for neighboring organizations; IAEA 18.4% and UNIDO 24.0% and the UN Secretariat 37.4%. Efforts by the PTS to increase the number of women are hampered by low numbers of female applicants for the majority of vacancies for scientific posts. The comparatively low numbers for IAEA and UNIDO might further illustrate this situation, and national scientific organizations also display similarly low numbers.

States and the PTS cooperate in training people through workshops and training courses (Fig. 10.2). These activities have different purposes. One such activity is to arrange fairly general courses to increase awareness of the Treaty among diplomats and technical experts. PTS has also organized specialized training courses for national experts on how to operate a monitoring station or a National Data Center and make use of the data and products provided by the PTS. A number of training courses in data analysis have also been conducted, preparing individuals to take up positions as data analysts at the IDC. A large number of experts from all over the world have also been trained as part of the development and testing of a training program for future inspectors for the on-site inspection regime. Over the first ten years a total of

Fig. 10.2a National Seminar on CTBT in Zambia February 2006

Fig. 10.2b National Seminar on CTBT in Kuala Lumpur, Malaysia 31 May-2 June 2007

3270 have participated in the workshops or training courses conducted by the PTS in cooperation with States Signatories. Table 10.3 gives the distribution of participants among the different regional groups and shows that the PTS and States Signatories have been quite successful in attaining widespread participation.

Fig. 10.2c International Cooperation Workshop for the region South East Asia, the Pacific and the Far East in Beijing, China, June 2000

Fig. 10.2d Workshop on on-site inspections in Hiroshima, Japan, July 2003. The picture is taken at the site of the memorial cenotaph for the A-bomb victims

10.1.4 States are PTS Customers

States receive data and analysis products from the PTS. These data and products are essential for the States to support the development of analysis procedures at their National Data Centers. The PTS also supports States in personnel training and by providing software, and in some cases hardware, to National Data Centers. The National Data Centers are on the other hand helping the PTS to evaluate the products supplied and the results of specific tests that have been conducted. This feedback is essential to improve the quality of the PTS products, as discussed in Chapter 8.

After entry into force, any concerns regarding compliance with the Treaty must be raised by States, and the CTBTO products will be an important basis for their monitoring of the CTBT. States need to work with the PTS prior to entry into force to ascertain they get the products they may need. They also have to establish competence and facilities, on their own or together with others, needed to interpret, in terms of Treaty monitoring, the information provided by the CTBTO after entry into force.

10.1.5 States are the Hosts of Monitoring Stations and Radionuclide Laboratories

Each of the 321 IMS stations has a host nation. The PTS needs to cooperate with these host nations in the establishment and provisional operation of the stations. A station also becomes the property of the hosting State once it has been established. The 16 radionuclide laboratories are national establishments of the host country and are paid a fee for service. The establishment of a station might be a long and at times complex process, as discussed earlier. States may play a number of roles in this process. One government or local authority might have to grant the necessary permits for the installation and another might have to authorize the data transfer needed, and agreements have to be reached with landowners. Institutions, governmental or non-governmental, or commercial companies in the host country are the ones carrying out the practical work of installing and operating stations. This work is performed under contract with the PTS on commercial terms. An IMS station might be part of a large organization or a national network that is also operated for other purposes, such as seismic hazard assessment. This might make it difficult to identify the specific resources needed for the CTBT-related activities. The logistical conditions might differ considerably from station to station, as might the conditions on the labor market.

To handle this complex situation, a large number of agreements and contracts have to be concluded with authorities, institutions and companies around the world. This process, dealing with States having different legal systems and traditions, has proved to be complex and time consuming. The context of an international treaty appears to have encouraged the PTS to take an overly formalistic, legalistic approach, even to purely technical and commercial issues. The internal PTS working procedure for handling contracts has also been complex and time consuming. States also share the responsibility for the delays as they often take a long time to respond to requests for proposals and to settle technical issues with the PTS. For these reasons and despite efforts to improve on it, the contracting process has, over the years proved to be a serious bottleneck and has slowed down the implementation of the IMS.

There is an added complication: a number of the experts participating in sessions of PrepCom's Policy Making Organs, in particular Working Group B, are senior staff members of national institutions hosting IMS facilities. In the elaboration in WGB they might easily run into difficulties due to conflicting interests. To increase awareness of this problem, the PrepCom plenary has decided that WGB participants should declare such interest to the WGB chairperson.

10.2   The CTBTO Preparatory Commission and International Organizations

10.2.1   Organizations in Vienna

During the negotiations suggestions were put forward that the verification task might be entrusted to the IAEA rather than creating a new organization. The decision was finally to establish an independent organization. The Treaty states that the organization "*shall seek to utilize existing expertise and facilities, as appropriate, and to maximize cost efficiencies, through cooperative arrangements with other international organization such as the International Atomic Energy Agency*". One of the important arguments for choosing Vienna as the site for the organization was the existence there of a number of international organizations, in particular the IAEA. PrepCom and its PTS took office in the Vienna International Centre (VIC), the home of several international organizations including the IAEA. So how has the cooperation developed? We think the experience is at best mixed.

PTS has participated in a close cooperation, already established among the other VIC organizations, on a number of practical issues that are common to all residents in the VIC. This includes large-scale conference facilities, security, health care, restaurants, office maintenance and cleaning. To be part of these joint efforts has proved cost-effective and the cooperation has been running smoothly.

On the other hand there has been no cooperation between the PTS and IAEA on matters of substance, despite the fact that there are several areas where the two organizations might benefit from sharing knowledge and experience, including radionuclide observations and on-site inspections. There might be different reasons why such a cooperation has not been developed. One is political; some States do not want to see any merger of activities related to the NPT and the CTBT and are thus reluctant to see any engagement by the IAEA in CTBTO Preparatory Commission activities, in particular in the radionuclide area, and vice versa. Another relates to relations among the VIC organizations on a more practical level and the difficulties of establishing cost-effective, smooth cooperation on activities such as libraries and personnel administration. Legitimate as these difficulties may be, they run counter to the ambition in the Treaty to develop cooperative arrangements. They have even prevented normal professional contacts between experts from IAEA and the PrepCom. A closer professional cooperation would no doubt be mutually beneficial to the two organizations and create a more attractive environment for the staffs of both organizations.

10.2.2 *International Humanitarian Organizations*

The International Monitoring System provides data and analysis results that are of great interest in relation to the mitigation of some humanitarian disasters. Even during the current period of testing and provisional operation, the system is operating with high reliability and can provide data in a timely fashion. The system thus has a great potential for being useful for a number of humanitarian applications (Massinon 2004, CTBTO Preparatory Commission 2007b). States have organized international seminars to identify such applications. Some

States, however, have been hesitant to allow the full use of data for humanitarian purposes. There seem to be two main reasons in principle. One is that data should remain classified and used only by States Signatories. Given that the observations of interest in this context relate to disastrous geophysical events or environmental conditions that have no relevance to Treaty monitoring, their release should present no security concern to any State. The other reason put forward is that the system should not be operational prior to entry into force of the Treaty, and no activities that could give the impression of an operational system should be permitted. It is clear from the resolution that established the PrepCom that no operational verification regime should exist before entry into force. There is, however, as we have discussed extensively, a need to test and evaluate the system and to operate it provisionally for those purposes. The provisional operation is conducted with relaxed operational requirements but even data provided in this non-operational mode would be most valuable for a number of humanitarian purposes and organizations.

Gradually, however, States have become more willing to allow the release of IMS data for humanitarian purposes. The tragic tsunami disaster in the Indian Ocean in December of 2004 caused many to understand that any data that could prevent or reduce the effect of future disastrous events should be fully utilized. Relevant data from the IMS network are now transmitted without delay to several tsunami warning centers. The IMS data have proved most useful to these centers, supporting their assessment of which earthquakes might present a tsunami risk.

10.2.3 Other International Organizations and Civil Applications

The CTBTO Preparatory Commission and the PTS have cooperative agreements with several international organizations. An important agreement is with the United Nations, adopted by the UN General Assembly on 15 June 2000 (CTBTO UNGA 2000). In a general way this agreement regulates the relations between the UN and the PrepCom, and it is an important recognition of the PrepCom as an international organization. More general political agreements have been concluded with the Agency for the Prohibition of Nuclear Weapons in Latin America and the Caribbean (OPANAL), signed in 2002, and with the Association of Caribbean States (ACS) in 2005. Despite those agreements, many States in those regions have not paid their assessed contributions. Since 2000 the PrepCom has had an agreement with the United Nations Development Program (UNDP). Under this agreement the PrepCom can get operational support services from the UNDP in dealing with local authorities in countries where UNDP is active.

Since 2003 PTS has had agreements on mutual cooperation on atmospheric modelling with the World Meteorological Organization (WMO) and the

European Centre for Medium-Range Weather Forecasts (ECMWF). The purpose is to estimate the movements of air masses to locate radionuclide sources from observations of radioactive particles or gases. This is of some mutual benefit as weather observations from IMS stations regularly flow to the WMO. In a longer-term perspective, when sources of radionuclide emissions are well known they can be used to tune and improve atmospheric transport models. Good models of the atmosphere are also a prerequisite for locating the sources of infrasound observations. Such observations may be used to detect volcanic eruptions at remote places to warn air traffic of ash clouds (see also Fig. 2.10).

Increasing concern about our global environment has made pollution monitoring on a global scale an increasingly important issue. The International Monitoring System has a global reach and might provide an interesting infrastructure to be used also for other measurements by adding the proper sensors. To use the local infrastructure at IMS stations, the data communications channels and the data collection facilities at the IDC could prove not only to be a cost effective way to establish a global pollution monitoring system, it is no doubt also the speediest way of implementing such a system.

Data already collected might prove useful for pollution monitoring. The particle filters to be collected routinely at 80 radionuclide stations around the world also collect a lot of non-radionuclide particles, carrying information on global pollution. Just looking at the particle filters collected routinely at such stations reveals a great difference between different parts of the world. Some filters are black from dirt while others look quite clean. Using new techniques it is today possible to make comprehensive chemical analysis, even of small particles. The filters collected today may thus carry most interesting information on global pollution, see Fig. 10.3.

A politically more controversial issue is to use the radionuclide data, both of particulates and noble gases, for detecting radioactive leakages of different kinds and origins. This use might be politically sensitive as it can provide information on ongoing nuclear activities of different kinds, including military purposes. Turning that argument around, the PTS radionuclide data might prove useful in strengthening the verification of the NPT. Perhaps the most useful might be to add some other functionality, like surveillance of krypton-85, another noble gas radionuclide, and basically use the IMS infrastructure as mentioned earlier. This would broaden to a global reach the wide-area radionuclide monitoring in the Additional Protocols of the NPT. It would most likely take a lot of political flexibility before such a synergy between the two treaties could be established, even if both treaties have a common goal of promoting nuclear disarmament and non-proliferation. It goes without saying that the IMS radionuclide data would be extremely useful for early mapping of the situation after a Chernobyl size reactor accident.

Fig. 10.3a Filter from radionuclide station RN08 at Cocos Island, Australia. The filters from this island site station are usually very clean

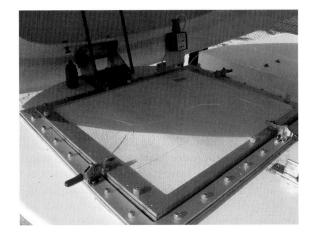

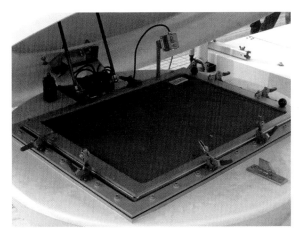

Fig. 10.3b Filter from radionuclide station RN20 at Beijing, China. The filters from this station are usually black and dirty

10.3 Synergy with Science

Science and technology played a crucial role in the design and development of the CTBT verification regime. The Treaty was developed with strong support from the scientific community and world famous scientists have also contributed to the negotiations and the implementation of the Treaty in different functions. The system was designed based on the scientific knowledge and the technologies available in the 1980 s and early 1990 s, when the S&T basis for the Treaty was created. The design of the system remains by and large unchanged. After 10 years of PrepCom activities, some of the equipment installed in the early years is nearing the end of its life cycle, and the PrepCom has now started to budget for recapitalization costs. In this regard, the PTS has estimated that the replacement value of equipment installed at IMS stations is about $62 million, and that the average lifetime of such equipment is 15 years. A number of technical components have already been updated as new technology has become available, and this is in particular true for the computer and communication systems and commercially available software. Key seismic data analysis methods used at the IDC, on the other hand, have not changed since the testing of prototype systems during the GSE work in the 1980 s and the first part of the 1990 s. As the seismic system could be provisionally operated quite early and a much appreciated bulletin was routinely produced, there were limited incentives to introduce major revisions that would affect routine bulletin production. The radionuclide technology, including atmospheric transport analysis, which was introduced in the CTBT negotiations at a quite late stage, has been modernized and developed much further in the first decade of the PTS. The same is partly true for the hydroacoustic and infrasound technologies.

Many States spent a lot of effort and money on national research and development projects to support this development, both prior to and during the CTBT negotiations. Once the Treaty was signed, many of those programs were cancelled or substantially reduced. This may be understandable, but it has a number of serious consequences. One is that many States are losing their ability to make an independent analysis and assessment of data to be provided by the CTBTO, after entry into force, as a basis for political decisions. Another is that it has become difficult for States to provide scientific support to the PrepCom and the PTS to maintain and update the verification system. It has finally reduced the ability of States to support the establishment of national facilities and competence in developing States to make CTBT verification a global undertaking.

The Treaty recognizes the need to incorporate future scientific and technological developments into the verification regime. Article VIII, on review of the Treaty, states that such review shall take into account any scientific and technological developments relevant to the Treaty. In the Protocol to the CTBT the general provisions for the IMS and IDC makes reference to *"any future agreed modification or development of the International Monitoring System"*. The Treaty also includes the possibility, after entry into force, to establish a Scientific Advisory Board.

With the present overly long PrepCom period there is also a need to recapitalize knowledge before entry into force. To be a viable, cost-effective organization, able to provide products of high quality, PTS has to stay in close touch with the scientific community and benefit from ongoing technical and scientific developments. The review of the PTS organization, referred to in Chapter 9, notes that *"scientific knowledge is evolving and methodologies used in verification need to be regularly updated in order to provide benefits to States Signatories. To achieve the full benefits of the investment in monitoring and analysis, there should be an integrated, coordinated and ongoing scientific development programme. Over time, increased efficiencies could be expected through automation"*. The review of the IMS program expressed similar views: *"for the continued integrity of the monitoring system it is essential to maintain close links with the scientific community"*. Advances in sensor technology, communications, computer hardware and software have been gradually implemented in the system. The future operation, maintenance and technical upgrades of the verification system require that new S&T developments be implemented to an even greater extent.

Knowledge too needs to be recapitalized. Core functions in the verification-related activities of the PTS rely and lean heavily on knowledge accumulated over a long period of time. Knowledge resides with the individuals working in the PTS, and these individuals and their knowledge are the main assets of the PTS. But people come and go, especially in a non-career organization like the PTS, so knowledge must be passed on to new employees, who also come with new ideas. Much of the knowledge of the PTS is represented and documented in procedures, methods, algorithms and computer programs, which have been developed, modified and enhanced by former and current PTS employees, and by members of the international scientific community. It is the task of PTS employees to ascertain that the equipment, methods and procedures used in the PTS take advantage of general scientific and technological developments. To achieve this, the PTS must be receptive to new ideas that emerge in the science and technology fields of relevance to the monitoring mission of the PTS and put these ideas to tests. If this is not happening regularly, the PTS will stagnate in its development, and the entire verification machinery may become a dinosaur with little promise for the future. Such a system will not fulfil its long-term goal of providing States Signatories with high quality products. States may also lose confidence in a system that the scientific community deems to be outdated.

Scientific and technical development relevant to strengthening security and non-proliferation is occurring at universities and research organizations around the world in a large number of fields. Many of these fields and activities are generic and not specifically tailored to test ban verification, but could still prove most important in keeping the verification and its products up-to-date. Over the years a number of cooperative projects have also been conducted between the PTS and scientific institutions to improve station design, analysis methods and computer programs. It is essential to develop this cooperation further to keep the verification regime up-to-date.

Many remaining issues related to the analysis of observations are crucial for States Signatories and the PTS. There is a need to establish a more cost-efficient analysis to cope with a growing flow of data as more and more stations are implemented. There is also the need to improve the quality of the products delivered to States Signatories. A particular problem is to more fully describe the events that have been observed to facilitate their interpretation by States Signatories. Many of these issues relate to the dramatic developments in infor-mation analysis at many scientific institutions around the globe. We are con-vinced that today's data analysis procedures could be most significantly improved by applying these new scientific developments.

This would afford not only a more cost effective analysis process, but also improve on the quality of the products to be delivered to States Signatories after entry into force of the Treaty. Application of the scientific developments to the analysis of verification data will also provide States Signatories with improved tools to analyze and interpret data provided by the CTBTO after the Treaty enters into force. To study how to apply new scientific developments for the analysis of CTBT verification data in an operational system, PrepCom should create an IDC development environment where new procedures and other scientific results can be implemented, integrated and tested. This should be an off-line system, open to the scientific community, using sets of IMS data for test and evaluation purposes.

There is a need to move further and develop a mutually beneficial, long-term cooperation between PrepCom and the scientific community. Data that are provided by the IMS network, even during the test and evaluation period, are of great value for scientific studies and scientific institutions. The PTS is applying S&T across scientific disciplines. This identifies problems that could stimulate new scientific developments of interest to the scientific community. This has already happened in the field of infrasound, where the data produced by the infrasound station network and the problems encountered in analyzing these data have re-established the corresponding scientific area.

A political issue related to the use of PTS data for scientific research emerged early during PrepCom. One State held the firm view that all data should, as a matter of principle, be available to any scientific institution without delay. Another State was equally firm in its view that there should be a certain delay in providing the data, to give States an opportunity to analyze what may have happened. This issue is still unresolved and has caused a serious, and in our view unnecessary setback in the relation between the PrepCom and the scientific community. From a practical point of view, data for scientific research are, in our view, normally not needed on-line and a certain delay should have no adverse effect.

To connect to the scientific community in a broader sense, a symposium entitled "CTBT: Synergy with Science 1996–2006 and Beyond" was held by the PTS in the fall of 2006 (CTBTO Preparatory Commission 2006). One of the themes at that symposium was data mining, with presentations of the recent dramatic developments that form the basis for a large number of web applications. It is clear that these developments could have a profound influence on how to improve the IDC data analysis (Fig. 10.4).

The PTS is not a research organization, but it is an organization that has benefited, and should benefit in the future from S&T developments in the scientific community. It is an organization that should integrate and apply new S&T developments. This means that the PTS not only has to stay in close touch with the scientific community but also needs to have mechanisms to evaluate and integrate developments that will improve the cost efficiency of the verification system and the quality of its products. Such mechanisms need to be applied in a systematic and sustained way. This is a question of technical facilities, such as the IDC development environment mentioned earlier, and of resources and procedures to promote such cooperation. It is also, and maybe most importantly, a matter of attitude of the PTS leadership and staff towards the scientific community. The organizational review and those of the IMS and the IDC programs all identified the need for a structured, long-term perspective on the PTS relations

Fig. 10.4 Usama Fayyad, Chief Data Officer and Executive Vice President, Research and Strategic Data Solutions, Yahoo! during the conference "Synergy with Science" in 2006, introducing the concept of data mining

with the scientific community. The organizational review states "*Rather than being insular and having inadequate contact with cutting edge technologies the CTBTO should become a relatively open organization that cooperates with other organizations*". The review further "*recommends that an integrated scientific development programme for the verification functions be undertaken to gain full benefits of past and future investments in monitoring and analysis*". The IDC review recommends that "*the PTS adopts a process of an ongoing scientific advisory function to help guide the development of the programme*".

In our view the PrepCom and the PTS must adopt a strategic approach to relations with the scientific community, in the same way as any other organizations or companies that depend critically on high technology and advanced science. To maintain and develop a viable organization and be able to deliver high quality products, PTS has to ensure the continuous influx and application of new knowledge and technology. To develop close, sustained ties with the scientific community is not a luxury that the PrepCom can choose to indulge or not as it pleases, or handle ad hoc; it is a truly strategic relation. Activities to explore and assess ongoing scientific and technological developments and to implement those that improve on the PTS operation must be an important task for the PTS, and for the Technical Secretariat after entry into force of the Treaty. The Policy Making Organs also have an important responsibility to make sure that this recapitalization of knowledge is an integral part of the PTS work program and that its implementation is carefully monitored.

To maintain and develop the CTBT as a global treaty is not only a question of developing the technical infrastructure and the methods used at the PTS for data analysis but also a question of capacity building in States Signatories. So far we have successfully connected stations and instruments around the world. Now it is time to connect people and their institutions. Through international scientific cooperation on a regional and global scale we need to develop the knowledge base and the facilities needed for States to be able to participate fully in the implementation and monitoring of the Treaty. Such cooperation will also enable States to benefit from the technologies involved in the verification system, for civil applications, too.

10.4 The CTBT – Hostage to Today's Politics

The CTBT is a political commitment and as such subject to the ups and downs of the political situation globally and in individual States. The most significant political setback to the Treaty was on October 13, 1999, when the US Senate voted against ratification of the Treaty. This has blocked any prospect for entry into force since then. Even if there are eight more States whose ratifications are needed for entry into force, the first move has to be made by the USA. The Treaty was derailed in Washington and it has to be put back on track in Washington. The USA showed strong leadership when the Treaty was negotiated; US leadership will also be needed to crack some of the remaining hard nuts to get the Treaty into force.

In many ways the Treaty gained strong support so far during the PrepCom period. States have generally cooperated constructively to implement the IMS stations around the world. The cooperation among States in the PrepCom has also by and large been constructive. States have supported the PTS by qualified personnel, and the PTS should be proud of having many competent staff members. Nevertheless, we have the feeling that political considerations sometimes are playing too important a role when hiring personnel, at the expense of qualification and experience. Hiring personnel is a critical process for any organization, institution or company; doing it in an international and political environment makes it even more difficult. In the case of the PTS, it requires great care by the PTS leadership and by States Signatories to balance the different interests that may be at stake.

Many States, close to a quarter of the States Signatories, are not paying their contributions and many never have done so. The financial consequences of this non-payment were initially limited, as most of the 42 States that have never paid are developing States with minimum assessed contributions. Now, in late 2007, the financial situation for PrepCom is most difficult. This is because the largest contributor, the USA, is not paying its contribution in full and because arrears are accumulating for States like Argentina, Brazil, Colombia and Iran. One result of this non-payment is that a large number of States have lost their voting right, which to some came as an unpleasant surprise when a new Executive Secretary was to be elected in the fall of 2004. More serious is the fact that a large number of States, by not paying their assessed contributions, demonstrate a lack of political will to live up to their international obligations.

The financial situation in 2007 illustrates the financial vulnerability of PrepCom/PTS and other international organizations, dependent as they are on a few States for the lion's share of their funding. Today five states – France, Germany, Japan, the UK and the USA – are supposed to pay more than 60% of the budget. This creates a strong political dependence on a few States, which one might want to avoid in an international organization. As the US decision not to pay its full assessment illustrates, one single State can bring an organization into severe financial difficulties. Contributions to international organizations established to promote our common security – IAEA, the CTBTO Preparatory Commission and OPCW – are, from a national security perspective, small amounts of money for any State. There is thus no reason why any State should refuse to pay its assessed contribution in full.

There might also be good reasons to consider putting a cap on the contributions from any one State, to limit the vulnerability. The cap should be set so that several States reach that level. This will even out the level of contributions among a larger number of economically strong States, without increasing the burden for those that today pay a minimum amount. More importantly, it will send a clear political message that the international disarmament, arm control and security building agenda does not solely depend on one or a few States, but is the responsibility of all.

At the PrepCom plenary sessions arguments are frequently heard, especially from the G77 Group, that the pace at which the verification system is built up should take account of the prospects for entry into force of the Treaty. This is usually used to argue for reduction of the yearly PTS budget – a budget to which many of these States have so far not contributed. The task of the PrepCom is to prepare for the entry into force of the CTBT and a most important element here is the delivery of an operational verification regime. We all wish that there were clear signs that entry into force was approaching, but there are none. This, in our view, must not lead to the conclusion that we shall reduce our efforts to implement the verification regime and proceed with the necessary tests and evaluations. Politics can change rapidly and we must not end up in a situation in which the PrepCom fails to deliver. Given that the PrepCom has been in existence for more than ten years and has thus been given more than ample time to conclude its task, it would be totally unacceptable if the verification regime were not ready at entry into force of the Treaty.

Chapter 11
Always Too Early to Give Up

Alva Myrdal, the late Swedish cabinet minister for disarmament, Nobel Peace Prize Laureate and also a strong test ban advocate, used to say "*it is always too early to give up*". Efforts for half a century to ban nuclear testing before the CTBT was finally signed in 1996 proved how right she was. This is by far the longest period that any treaty has been in the waiting room. A number of initiatives were taken during this time. Many individuals, delegations, institutions and States throughout the years devoted a lot of effort, ingenuity and money to paving the way for the Treaty. The path to the Treaty is marked by optimism and progress, disappointments and failures. The intent of this chapter is to capture lessons learned from the negotiations, the Treaty itself and its implementation.

We do this with the intent of facilitating the never-ending process of keeping the CTBT a viable treaty, a process that strengthens the support of the Treaty and above all makes the Treaty enter into force, a process that is able to fully utilize the ongoing scientific and technological developments in cooperation with the scientific community to provide all States Signatories with verification data and services of increasingly higher quality, to enhance the ability of all States to be equal partners in the implementation of the Treaty, and finally, a process that maintains the CTBTO Preparatory Commission as an efficient organization that will continue to attract competent people.

Treaties are of necessity negotiated one at a time, taking account of the issue at hand and the actual political situation. Experience has in rare cases been carried from one negotiation or one treaty to the next, mainly by persons who have contributed to more than one negotiation. We are convinced that there is a lot to be learned from past experiences by a more thorough and documented analysis of past negotiations, of the treaties themselves and their implementation. We hope that our analysis of experiences from the long history of the CTBT will prove useful when negotiating and implementing new treaties.

In the introduction to the book we stated that we would attempt to analyze the negotiations of the CTBT, the Treaty itself and the efforts to implement it from three different perspectives: political, scientific and managerial. We have done so at a number of places throughout the book and in the first part of this last chapter we summarize our assessments under these three headings.

O. Dahlman et al., *Nuclear Test Ban*, DOI 10.1007/978-1-4020-6885-0_11,   223
© Springer Science+Business Media B.V. 2009

We also reflect on the prospects of bringing the Treaty back on track. Finally, we briefly discuss the new broad security agenda where it is non-military crises that are the main threats to our societies. We argue the need to use science and technology to cope with these threats and to invest in non-military security in the same way as we have traditionally invested in military security.

11.1 Finally a CTBT – and Then? A Political Perspective

International treaties and agreements are playing an increasingly vital role in promoting international security in an environment where non-military elements are becoming more important. Today, efforts to preserve the environment and prevent the proliferation of weapons of mass destruction are at the forefront of the international security agenda. The signature of the CTBT, fifty years after the first initiatives to limit nuclear testing were taken, was an important political achievement and it was rightly hailed as a success. Not only is the CTBT an essential element in a global non-proliferation regime and a step towards nuclear disarmament, the conclusion of the Treaty also removed a hurdle en route to further nuclear arms control and disarmament treaties. As we now know, further obstacles have emerged on that route.

The long process needed to achieve the CTBT illustrates that progress in international negotiations is generally slow and held hostage by the general political situation, especially among the key players. It also shows that one should never give up but always keep trying and that partial measures, imperfect as they might be, could be important stepping stones. A step-by-step process might stand a better chance of leading to the desired goal than an all-or-nothing approach. Such a step-by-step process might itself build the mutual trust and confidence necessary for a final agreement.

The CTBT was successfully negotiated within the framework of the Conference on Disarmament. The CTBT negotiation is rightly regarded as a major achievement for the CD, but it also clearly demonstrated its inherent weakness, further amplified by ten years of inaction following the CTBT. CD is held hostage by its size, its need for consensus on any issue and its bureaucratic and traditional proceedings. The need to reach consensus even to start negotiating or discussing a particular issue or to initiate exploratory scientific and technical studies, has paralyzed international disarmament and arms control work for a decade. We all know that in the end those who join a treaty have to agree, but we should find new ways to explore different issues to pave the way for serious negotiations. New processes have to be found, which do not carry the burden of achieving consensus in each and every step. CD was a valuable forum for dialog during the Cold War period, but if it is to move forward with a number of issues related to the new security agenda, the world today needs a vigorous, flexible negotiating environment. This could be a modernized CD or it could be environments where the negotiating forums are created to fit the issues at hand.

The signing of the Treaty marked the end of a very long journey, but it also proved to be the beginning of another. The impetus to international disarmament efforts created by the agreement on the CTBT has been gradually fading away during this long PrepCom period. The achievement of a norm against nuclear weapons testing that is most difficult for any State to break, is partly overshadowed by the failure to bring the Treaty into force. Given the present experience that, despite 178 signatures and 144 ratifications by February 2008, (see Fig. 11.1) the Treaty has not yet entered into force, it might have been more attractive to follow a more conventional entry into force (EIF) procedure that would have brought the CTBT into force long ago. The ambition to involve all key States from the beginning is a most respectable one but it has to be weighed against a long or very long delay in the entry into force. The CTBT has proved that you take a great risk when you create a mechanism that makes it possible for a single State, or in the case of CTBT nine States, to hold a treaty hostage. An early EIF might also have created stronger pressure on States that so far have not signed or ratified the Treaty. On the other hand, we have to acknowledge that full implementation of the Treaty requires the political will and commitment of all States, whatever the technical arrangements for EIF may be.

The CTBT has an unprecedented international verification regime. The aim of these verification measures is to give all States Parties an equal and fair possibility to monitor compliance with the Treaty by providing high quality information to all States. The CTBTO will be a service organization tasked with collecting, compiling and making routine analyses of all available data and providing all the data and information to States Parties. This is most valuable support as few States have the ability or resources to carry out this extensive work on their own. To be able to agree on such an extensive, complex verification system, and to implement it on a global scale, is a strong confidence building measure. It further demonstrates to those that will negotiate future treaties that complex, extensive verification systems can be made integral parts of such treaties and that such systems will give States more equal ability to verify compliance.

The CTBTO will not, however, have the authority to draw conclusions regarding the nature of the events that are observed. Although the Treaty text is not very explicit, it is unequivocal that it is up to the States Parties to assess compliance with the CTBT and to request further actions such as consultations and on-site inspections. To do this States need to establish competence and facilities, on their own or together with others, to interpret the information provided by the CTBTO in terms of treaty monitoring. We have noted that many States have reduced their capacity and competence in test ban verification over the last ten years, and few States are today able to make independent assessments of Treaty compliance. To make the Treaty and its verification regime a truly global undertaking it is thus necessary to increase engagement

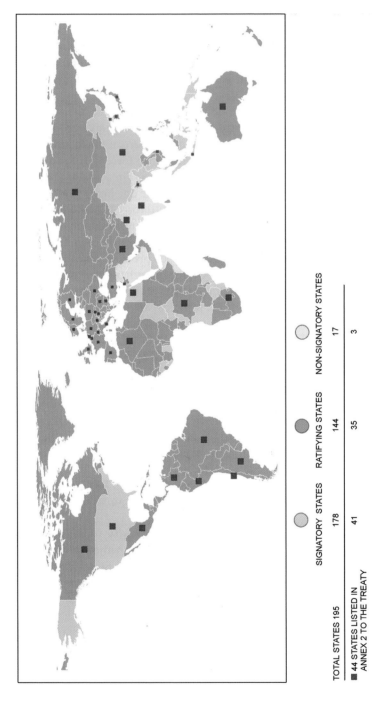

Fig. 11.1 Status of CTBT signatures and ratifications as of February 2008

among States Signatories to ascertain that the competence and resources needed are available at entry into force.

During a long PrepCom period States have cooperated well and supported the implementation of the Treaty. During many years they have also provided PTS with generous budgets. This good financial situation for a long time masked the fact that 42 States, most of them with the minimum assessed contribution, have never paid anything. Over the years the collection rate has been declining and the PrepCom's economic rules enhance that problem. The delayed US payments in recent years and the accumulated non-payment from a number of States – Argentina, Brazil, Colombia and Iran in particular – have amplified the situation and caused the PTS to face a financial problem. By not paying their assessed contributions in full, almost half of the States Signatories are not living up to their international obligations. This is a severe blow to the global engagement in the promotion of nuclear disarmament and non-proliferation.

The difficulties for the PrepCom and its PTS caused by the non- or limited payment by a number of States Signatories are an illustration of the vulnerability of international treaties and organizations to actions by individual States. The non-EIF of the CTBT and the difficulties facing the NPT by nuclear weapon States outside the Treaty can be seen as other examples. It thus seems important to address the resilience of international treaties and organizations and to take measures to make them less vulnerable to actions by individual States. Making international organizations less dependent on large contributions by one or two States could be one such measure.

11.2 Most Complex Verification System Ever – a Scientific Perspective

The CTBT proves that it is possible to reach agreement on a highly technical, complex verification regime with a global reach, which also provides for most intrusive on-site inspections. The PrepCom period has proved it possible to implement this extensive, complex international monitoring system all around the globe. The implementation has taken much longer than initially expected. The reason is partly political, as there has been no pressure to speed up the implementation when the entry into force is not in sight. The reason is also partly bureaucratic and technical. The efforts and time needed were initially greatly underestimated both for obtaining the necessary permissions to establish stations in 89 States around the globe and for the actual technical build-up of the stations. The cost of the system was also greatly underestimated by the negotiators in Geneva, by the PrepCom and initially also by the PTS. These difficulties in getting reasonable estimates of efforts, time and money are in no way unique to the CTBT; they are experienced in the establishment of most large-scale, complex systems. These difficulties, however, reveal the need to engage system engineers and project managers at an early stage of the elaborations.

The scientific and technical preparatory work carried out by the CD Group of Scientific Experts greatly facilitated the negotiation and the implementation of the CTBT. This method of work could be applied to other international security issues. It should be possible to conduct such work in ways that do not imply any commitments that political negotiations might follow or in any way prejudge their possible outcome. The results of such technical and scientific elaborations might serve as a useful background to any political negotiation that might follow. Joint international research and development projects might be part of such preparatory work and States might be encouraged to establish infrastructure in support of such projects. As in the case of CTBT, verification arrangements might also in future prove to be both extensive and complex and would thus not be an element that could be added at the end of a negotiation. A credible verification regime might, on the contrary, well be a prerequisite for an agreement. Such international scientific and technical cooperation could also help create mutual confidence. In our view we should aim at creating an international mechanism by which technical and other non-political issues, related to our security agenda, could be explored at an early stage. The Intergovernmental Panel on Climate Change, IPCC (IPCC 2007), which shared the Nobel Peace Prize for 2007, is a good example of such an extended scientific effort, on a global scale, to address a most important security issue.

As we discussed earlier, the CTBT enters into force six months after the last ratification of the 44 States listed in Annex 2 to the Treaty. The verification system available at that time will constitute the operational system. How complete will this initial system be? By the end of 2007, the hydroacoustic network and large parts of the seismological and radionuclide networks and almost two thirds of the infrasound network (39 of 60 stations) were in place, and installation of remaining stations is progressing. The PTS needs to establish a workable readiness plan of how to finalize the remaining station installations, and host States have to cooperate fully to make that happen. Procedures are implemented at the PTS to analyze data, and bulletins containing the results of this analysis are provided regularly to a large number of institutions in States Signatories. Given that States provide the necessary funds and that PTS focuses its work on the remaining key elements of the system, we should by now have passed the time when there would be any uncertainty with respect to the readiness of the system at entry into force.

How capable will the system be? The system is not designed to meet any agreed performance goals. During the negotiations, States might have had their individual expectations of what the system might deliver. Prototype seismic systems were extensively tested during the work of the GSE, giving a fair basis for estimating its capability. The preliminary results so far available indicate that most likely many of those expectations will be exceeded. During the establishment the system has benefited from more than ten years of rapid development in science and technology. The dramatic scientific development in information technology offers a great potential to further upgrade the data analysis procedures and to improve the quality of the products provided to States. The verification system was regarded as adequate when the Treaty was signed. The developments since then have enhanced confidence in the verification system and its ability to provide data to allow States to adequately verify compliance with the CTBT.

The provision for on-site inspections is an essential and politically sensitive part of the verification regime. More time and frustration have been spent, during the PrepCom, on establishing instructions on how to conduct such inspections than on any other issue. Provisional instructions, in the form of a Test Manual, have now been agreed, to be used in a major OSI exercise in 2008. Based on that experience, it should be possible soon to finalize an OSI Operational Manual, given the political will of the States concerned.

We still need to develop the more general philosophy around OSI. We saw during the UNMOVIC inspections in Iraq how difficult it was to accept that you did not find the weapons of mass destruction you were looking for. How much inspection is enough? When do you finish an inspection if you do not find a clandestine activity or event? How do you conduct an on-site inspection in such a way that you can come out of the inspection with a higher degree of confidence than when you went in? An OSI under the CTBT, if one is ever conducted, may be used to clarify a naturally occurring event, most likely an earthquake. Earthquakes may not leave any discernible marks on the surface. How then do you end an inspection when you cannot find any positive evidence of either an explosion or an earthquake? There is a need for further thought on such questions, not only in relation to the CTBT but also in relation to on-site inspections more generally.

11.3 Challenge to Establish a Technical Organization in a Political Environment – a Managerial Perspective

To establish and operate an international, highly technical organization in a political environment is in many ways a challenge. The fact that the PrepCom by February 2008 has 178 States Signatories as its masters and conducts work in 89 countries around the globe adds to the difficulties. States and the PrepCom have a complex pattern of contacts and relations. States are the masters of the implementation and provide the funds and personnel needed. They are also the customers of the PTS products and the hosts of the stations in the monitoring system. Institutions and companies in States Signatories are also contractors to the PTS to build and operate stations. States and their representatives and the leadership of the PTS have, in our view, over the years handled these multitudinous relations in a fair and professional manner.

It is not obvious what constitutes a good balance of work between the PrepCom Policy Making Organs and the PTS, and the balance has also shifted over time. During the first few years, when everybody expected an early entry into force, there was very close cooperation to move issues along speedily and informally. Once it became evident that entry into force was remote, the role of each body became better defined. The ambition that the Policy Making Organs should set well-defined, annual goals for the PTS, and then follow up on those goals at the end of the year, was not fully realized. It proved hard to define the goals as well as to engage States and the PTS in a credible follow-up process.

Looking more specifically at the PTS from a managerial perspective, one might discern several elements: the structure and working practice, and issues directly related to the staff. The initial structure, or the organization of the PTS was decided at the first PrepCom session. That decision was, at least partly, guided by political rather than managerial considerations, as States and groups of States sought to gain influence by securing particular positions for themselves. We believe it is essential to base the organizational structure on managerial, and not primarily political considerations, and to create an organization that promotes efficiency and internal cooperation and ascertains that the necessary competence and experience can be combined at all levels of the organization.

The efficiency of an organization also depends on the rules and regulations that guide its work and on the established work practice. The rules and regulations that guide the PTS build closely on those of the UN. A tendency to be overly formalistic and bureaucratic may be unavoidable in international organizations closely linked to the political environment, which may lead to a less cost-effective, more rigid organization. The rules have created great problems for the PTS, both when it comes to economy and to the personnel. As there seem to be few, if any, assessments made of the consequences of the rules and regulations used in many international organizations, it might be timely to conduct such an evaluation on a broad scale, involving a number of international organizations.

To be in charge of a technical international organization such as the PTS is a multifaceted task, requiring professional leadership with a mastery of a high level political and diplomatic environment and a highly specialized scientific and technological organization. These are two quite different tasks, each of them demanding in its own right. It is unreasonable and unrealistic to expect that the skill and experience needed can be combined in a single person. It is therefore essential for organizations like the PTS, and there may be other highly technical international organizations facing the same challenge, to create a team at the top within which these skills can be combined. For the PTS a proposal was put forward when the organization was established to create an additional senior position in the office of the Executive Secretary, to support the ES by taking charge of the technical and internal managerial issues. This idea was rejected at that time. The PrepCom period has clearly shown the need for such a strengthening of the ES office and we strongly advocate that such position be created in the PTS and in the Technical Secretariat after EIF. Another important experience is that international organizations, in the same way as national ones, need staff members with good managerial skills and experience at all levels in the organization. The need might be even greater in an international organization, with a more heterogeneous staff composition. The need for managerial skills has to be seriously considered by both PTS and States in the future staffing of the organization.

We have witnessed, over the years, that individuals matter very much in international work. We have also seen the importance of creating trust and confidence among those sharing a common task, and to form teams with common goals. We saw this during the negotiations in Geneva, where some individuals had a most significant impact on the negotiations and where the negotiators eventually worked in a constructive team to find common solutions. We have also seen it in the PTS where, over the years, a number of persons have, formally or informally, taken the lead to move issues forward significantly. We have also seen examples of good team building as well as examples where teamwork has been absent. To hire people and to create a good team from a number of individuals from all over the globe is a difficult process. It requires an attitude of respect for the individuals and their particular skills and personalities, and sustained team building efforts. It also requires a humble insight that individuals are not directly interchangeable and that deliberately breaking up a functioning team has a great cost.

The PrepCom has two principal, equally important tasks: to deliver an operational verification regime at entry into force and to develop PTS into a well functioning organization that can be seamlessly transformed into the Technical Secretariat of the CTBTO. We have seen that, after a long build-up period, the verification regime is finally well underway. It is now essential also to revitalize the organization and make sure that it has the staff and the internal work methods in place to meet the challenges at entry into force. The PTS has recently been reorganized in a way that should facilitate a seamless transformation into the Technical Secretariat of the Organization at entry into force.

During the negotiations it was suggested that the IAEA should also be entrusted with the task of implementing the CTBT in addition to the safeguards agreements for the NPT. For different reasons this was not considered an acceptable solution. It was, however, argued that synergy with IAEA would be obtained by locating the PrepCom and the PTS in Vienna. Such synergy has so far been achieved only on a purely logistic level with the international organizations in the Vienna International Centre. Given the hurdles when creating new organizations, careful consideration should be given to the possibility of using existing organizations for new tasks. The IAEA existed at the time of the NPT negotiations and was given the additional task of verifying compliance with that treaty. The IAEA is also well placed to take on whatever verification measures that might be part of an eventual "cut-off" treaty.

The political body for a treaty, such as the CTBTO, has, for a number of political and other reasons, to be specific to that particular treaty. The question is; could such a political body outsource the bulk of the technical work to an existing organization? Put differently, could one and the same technical organization or secretariat manage the technical support for different treaties and thus serve several political masters? What would be the advantages and the disadvantages? Would this possibly be a more cost-effective way? And, maybe more importantly, would it speed up the implementation of a new treaty, given the large amount of time and effort needed to establish a new international organization able to conduct complex technical work? Could existing organizations be tasked with exploring new issues in the disarmament and arms control field to provide material and expertise that might be used in future negotiations? The present experiences from the negotiation and implementation of the CTBT are not providing answers to those questions, but they do make it relevant to put the questions.

11.4 Bringing the CTBT Back on Track

The CTBT has a built-in mechanism to promote entry into force. In Article XIV of the Treaty the depositary, the Secretary General of the UN, is mandated to convene a conference to promote EIF, if the Treaty has not entered into force three years after it was opened for signature. The first of those conferences was held in 1999 and the fifth in September 2007. The Article XIV Conferences have not been able to change the attitude of those States that are key to the EIF of the Treaty, and it is not likely that such a formal setting by itself will promote EIF in any significant way. There is a need for more focused, long-term work by States Signatories at political and diplomatic levels to convince remaining States of the benefits of signing and ratifying the Treaty. Such concerted efforts to bring remaining States on board have so far been absent.

So what are the prospects? Let us first discuss the situation in the USA, as it is key to the issue and provides a great deal of transparency on its elaborations. In his report (Shalikashvili 2001) to President Clinton on January 4, 2001, General J. Shalikashvili identified the following four most prominent concerns expressed in the debate leading up to the US Senate decision in 1999:

- Whether the CTBT has genuine non-proliferation value;
- Whether cheating could threaten US security;
- Whether USA can maintain the safety and reliability of the US nuclear deterrent without nuclear explosive testing;
- Whether it is wise to endorse a CTBT of indefinite duration.

After his review of the Treaty's impact on US national security, General Shalikashvili concludes "*I believe that an objective and thorough net assessment shows convincingly that US interests, as well as those of friends and allies, will be served by the Treaty's entry into force*".

Since General Shalikashvili's report was published, a number of comments have been made addressing the issues he identified. The US Nuclear Weapon Complex has repeatedly stated that it has proved possible to safely maintain the US nuclear arsenals since the testing stopped in 1992. A program, the Reliable Replacement Warhead (RRW), has been proposed to explore the option of developing a new warhead that is more secure, easier to maintain and can be certified without testing. We may not see the need for yet another generation of nuclear weapons, but the conclusion from this must be that it is possible to maintain credible nuclear deterrence without testing.

Another widely discussed issue is the monitoring of the Treaty: is it providing sufficiently adequate assurances that all parties fulfil their obligations? Our conclusion is that the verification regime will be more capable than most parties expected at the time the Treaty was signed. The main issues standing in the way of the entry into force of the CTBT are thus likely to be not technical but political ones. What are the prospects of a changed attitude? Are there any signs? Let us look at a few.

In an article in the Wall Street Journal on January 4, 2007 (Shultz et al. 2007), which attracted great attention, the two former Secretaries of State, George Shultz and Henry Kissinger, former Defence Secretary William Perry and former Senator Sam Nunn "*endorse setting the goal of a world free of nuclear weapons and working energetically on actions required to achieve that goal*". They propose a number of concrete steps including: "*Changing the Cold War posture of deployed nuclear weapons to increase warning time. Continue to reduce substantially the size of nuclear forces in all states that possess them. Halting the production of fissile material for weapons globally*". On the CTBT they propose "*initiating a bipartisan process with the Senate, including understandings to increase confidence and provide for periodic review, to achieve ratification of the Comprehensive Test Ban Treaty, taking advantage of recent technical advances, and working to secure ratification by other key states*".

Other powerful political voices are being heard calling for a world free of nuclear weapons. This was precisely the title of the keynote address by Margaret Beckett, then Secretary of State for Foreign and Commonwealth Affairs, UK, at the Carnegie Endowment Conference on June 25, 2007 (Beckett 2007). "*What we need*" she said "*is both a vision – a scenario for a world free of nuclear weapons. And action – progressive steps to reduce warhead numbers and to limit the role of nuclear weapons in security policy*". Referring to the NPT signed in 1968 she continued "*The judgement we made forty yeas ago, that the eventual abolition of nuclear weapons was in all of our interests- is just as true today as then. For more than sixty years, good management and good fortune have meant that nuclear arsenals have not been used. But we cannot rely on history just to repeat itself*". On the CTBT Ms Beckett stated "*We need to press on with both the Comprehensive Test Ban Treaty and with the Fissile Material Cut-Off Treaty. Both limit – in real and practical ways- the ability of states party to develop new weapons and to expand their nuclear capabilities. And as such they therefore both play a very powerful symbolic role too- they signal to the rest of the world that the race for more and bigger weapons is over and that the direction from now on will be down not up. That is why we are so keen for those countries that have not yet done so to ratify the CTBT. The moratorium observed by all nuclear weapon states is a great step forward; but allowing the CTBT to enter into force – and, of course, US ratification would provide a great deal of impetus – would be showing that this is a permanent decision, a permanent change in the right direction".*

In an article in February 2007 (Sterngold 2007), Dr Joseph Martz, leader of a team designing a new generation of warheads at the Los Alamos National Laboratory, argued that the key to the new US nuclear policy would be to slowly reduce the number of warheads over a period of years and, as an interim phase, replace older weapons with the new RRW. The final goal should be the elimination of the entire nuclear arsenal. What the USA should retain in its place would be the technology to assemble warheads from stockpiled material if a grave threat to national security were to arise. He also argues that USA could *"bolster its credibility as a leading voice for disarmament by ratifying the long-stalled treaty banning underground testing"*. US perspectives on the CTBT have recently been further discussed in several articles (Hafemeister 2007, Jeanloz 2007, Medalia 2008, O'Hanlon 2008).

How about some other key States? Three nuclear weapon States – France, Russia and UK – have ratified the CTBT, and thus do not see the need for any further testing to support their nuclear weapons programs. France has also completely dismantled her test sites. China has stated that the People's Congress is and has for a long time been considering ratification of the Treaty. China has not expressed any need or intent to resume testing. It would not be surprising if China would find it timely to rapidly ratify the CTBT as soon as the USA has done so. The process of disarming North Korea of its nuclear weapons capability is progressing with the closure of the Yongbyon reactor in mid-July 2007. It should be possible to get North Korea to adhere to the CTBT as part of that nuclear disarmament process.

At the end of the CTBT negotiations India was most dissatisfied with the result and opposed having the Treaty opened for signature. Many years have passed and both India and Pakistan have declared a moratorium on nuclear testing and further declared that they will not stand in the way of entry into force of the CTBT. It may not come easily, but it should be possible for the main nuclear powers, in particular the USA and China, to convince India and Pakistan that it is in their best interest to sign and ratify the CTBT.

In the Middle East, with Egypt, Israel and Iran, the general geopolitical situation is the main obstacle. As far as we know, Israel has not conducted any testing program to develop whatever nuclear weapons it may possess. It is concerned about the procedures for on-site inspections but should from other technical perspectives have no problem ratifying the CTBT. Egypt has no nuclear weapons program and should be able to ratify if Israel does. Iran is a party to the NPT and claims repeatedly that all its nuclear activities are for civilian purposes, in full compliance with the NPT. To ratify the CTBT would be a clear way to demonstrate its peaceful intent, and such a step might also facilitate the discussion on how to increase its international cooperation in the nuclear energy program.

The last of the nine current hold-out States – Indonesia – seems to have some internal hurdles to overcome to ratify the Treaty. Indonesia is a non-nuclear weapon State and party to the NPT, and it is unlikely that it will delay the entry into force of the Treaty.

11.5 A New Security Agenda

The security perspective of today and tomorrow is broader than that of yesterday. Globalization has also made us more mutually dependent. Crises far away can affect us all and security has to be built on a global scale. States can no longer create local safe heavens. As the mutual dependence increases and the threats become truly global, we need to expand the multilateral capability to cope with these threats. On the global scale we are moving from deterrence to confidence building, from armed conflicts to crisis prevention and management. Conflicts today occur within rather than between States and non-state actors are playing an increased role.

The security of each of us is increasingly coming to depend on our ability to handle non-military crises. The terrorist attacks in Spain, UK and USA are tragic illustrations of the vulnerability of our societies. We are also facing a growing number of threats to our modern societies of an economic and technical origin. Deliberate attacks on our information systems to manipulate the very nerves of our societies can be launched from any point on earth and by small groups of people. The non-military threats, in addition to terrorism, also include organized, more "conventional" crime, trafficking and drugs. In many parts of the world the greatest threats to the security of children, women and men are still famine and disease. The threats to our global environment are coming to the forefront of national and international political agendas.

Many of these threats confronting us are reflected every day on the front pages of our newspapers. We have, in short, moved from a situation where we were planning to cope with disastrous military confrontation that might occur with a low probability to a situation where we have to cope with a number of threats to our security and safety facing us every day. Proliferation of weapons of mass destruction is a particularly dangerous threat. To imagine a nuclear weapon in the hands of a terror organization is a most fearful scenario.

We are now facing a number of important and difficult questions:

- How do we increase the resilience of societies around the world to these new threats?
- How do we invest wisely in security to cope with this broader spectrum of threats?

- How do we proceed to effectively enhance our collective security and the security of individuals?
- Do we need to broaden the institutional tools at our disposal?
- How do we apply science and modern technology to this new and broad spectrum of threats to our security in a way similar to what we have done for centuries in the military area?

11.5.1 Science in Support of Security

The rapid developments of our societies over the last century are based on unprecedented achievements in science and technology. Science and technology have shaped our modern societies and have also been used extensively to enhance the capability of the military component of our security system. How can science and technology play an equally strong role in promoting security in this new, broad perspective? Not as a substitute for political will and action, but as a means to facilitate concrete results. How could we focus our attention on these problems in a similar way to what we have done so far on the military side? How could science and technology make a difference when it comes to confidence building, arms control, and international co-operation to increase the resilience of our societies to a growing number of threats? CTBT is a good example of the successful application of science and technology to global non-military security issues: how do we carry this good example further into new domains?

New scientific results and knowledge is created primarily within the individual disciplines and the challenges and rewards are also discipline oriented. Such work, including that also in a number of fields of importance to strengthening security, is being done at universities and research organizations around the world. Creating knowledge and effectively applying knowledge are two different processes and they need different environments in which to develop optimally. The application of knowledge is problem oriented and in most cases involves several disciplines.

It is reasonable to assume that we are able to utilize and apply only a small fraction of the scientific results and the knowledge created around the world. To create new knowledge within the different individual disciplines is time consuming and costly. To integrate and apply knowledge and experience created somewhere else is a most cost-effective process to harvest the results of efforts made by many over long times. Knowledge also has the nice quality of growing when it is shared. To more efficiently utilize the knowledge already created and available would be a cost-effective "force-multiplier" to our societies and in building global security. We need to create environments where issues related to the broad security agenda of today and tomorrow can be addressed in an integrated fashion.

11.5.2 Invest in Security

Ever since they were created, States have invested heavily in the military component of their security regimes. Until a few years ago security for a State was synonymous with a strong military defence. The security concept is different today and investments in security must follow suit. States individually and in international cooperation must now be prepared also to invest in non-military security building arrangements. So far such investments are negligible in comparison with what goes into the military forces. The combined annual budgets of the IAEA and the PrepCom, the two international organizations entrusted with the task of monitoring nuclear arms control, comprise in 2007 about $470 million. This is a small amount in comparison with the annual budget of $4.5 billion for the US Stewardship Program to maintain the readiness of the US nuclear weapons, and this is for one nuclear power only. It is also small in comparison to what small States spend on military defence: the Swedish assessed contribution to international organizations is about 1% or around $5 million for IAEA and the PrepCom combined, which is less than a tenth of a percent of its defence budget of around $7 billion.

The political attitude of States has to change. It must be politically unacceptable that States refuse to pay their assessed contributions to international organizations established to promote global security. There is a need for more activities and more contributions, not less. There might also be reasons to consider how States share the burden of financing international organizations in the fields of disarmament, arms control and security building. Today a few countries contribute the lion's share to such organizations. Five states – France, Germany, Japan, UK and USA – pay more than 60% of the budgets. This creates a strong political dependence on a few states, and a corresponding vulnerability, which should be avoided. As we are considering, in national terms, comparatively small amounts of money, there might be good reason to consider putting a cap on the contributions from any State. The cap should be set so that several States reach that level. This will even out the level of contributions among a larger number of economically strong states, without increasing the burden on those which today pay a minimum amount. More importantly, it will send a clear political message that the international disarmament, arm control and security building agenda is not entirely dependent on one or a few States, but is the responsibility of all.

Annex 1

This annex lists all 321 stations of the International Monitoring System, with their geographical coordinates as currently (late 2007) available. The rightmost column shows the deviations (*Dev*) from the Treaty locations. A dash (–) in this column indicates that the station has no coordinates listed in the Treaty, or that station installation has not started and the eventual station location has not been established. The annex also lists current information on the 16 radio-nuclide laboratories.

Primary Seismic Stations

Treaty No.	Station	State	Type	LAT	LON	Dev (km)
PS01	Paso Flores	Argentina	3C	40.7 S	70.6 W	0
PS02	Warramunga, NT	Australia	Array	19.9 S	134.3 E	0
PS03	Alice Springs, NT	Australia	Array	23.7 S	134.0 E	0
PS04	Stephens Creek, NSW	Australia	3C	31.9 S	141.6 E	0
PS05	Mawson, Antarctica	Australia	3C	67.6 S	62.9 E	0
PS06	La Paz	Bolivia	3C	16.3 S	68.1 W	0
PS07	Brasilia	Brazil	3C	15.6 S	48.0 W	0
PS08	Lac du Bonnet, Man.	Canada	3C	50.2 N	95.9 W	0
PS09	Yellowknife, N.W.T.	Canada	Array	62.5 N	114.6 W	0
PS10	Schefferville, Quebec	Canada	3C	54.8 N	66.8 W	0
PS11	Bangui	Central African Republic	3C	5.2 N	18.4 E	0
PS12	Hailar	China	3C	49.5 N	119.8 E	23
PS13	Lanzhou	China	3C	36.0 N	103.7 E	14
PS14	El Rosal	Colombia	3C	4.9 N	74.3 W	0
PS15	Dimbokro	Côte d'Ivoire	3C	6.7 N	4.9 W	0
PS16	Luxor	Egypt	Array	26.0 N	33.5 E	50
PS17	Lahti	Finland	Array	61.4 N	26.1 E	0
PS18	Tahiti	France	3C	17.6 S	149.6 W	0

(continued)

Treaty No.	Station	State	Type	LAT	LON	Dev (km)
PS19	Freyung	Germany	Array	48.8 N	13.7 E	11
PS20	TBD*	TBD	TBD			0
PS21	Tehran	Iran (Islamic Republic of)	3C	35.9 N	51.1 E	29
PS22	Matsushiro	Japan	Array	36.5 N	138.2 E	0
PS23	Makanchi	Kazakhstan	Array	46.8 N	82.3 E	23
PS24	Kilimambogo	Kenya	3C	1.1 S	37.3 E	11
PS25	Songino	Mongolia	Array	47.8 N	106.4 E	37
PS26	Torodi	Niger	Array	13.1 N	1.7 E	–
PS27	Hamar	Norway	Array	60.8 N	10.8 E	0
PS28	Karasjok	Norway	Array	69.5 N	25.5 E	0
PS29	Pari	Pakistan	Array	33.7 N	73.3 E	–
PS30	Villa Florida	Paraguay	3C	26.3 S	57.3 W	0
PS31	Wonju	Republic of Korea	Array	37.5 N	127.9 E	0
PS32	Khabaz	Russian Federation	3C	43.7 N	42.9 E	0
PS33	Zalesovo	Russian Federation	Array	53.9 N	84.8 E	0
PS34	Norilsk	Russian Federation	3C	69.3 N	87.5 E	39
PS35	Peleduy	Russian Federation	Array	59.6 N	112.6 E	0
PS36	Petropavlovsk-Kamchatskiy	Russian Federation	Array	53.1 N	157.7 E	7
PS37	Ussuriysk	Russian Federation	Array	44.2 N	132.0 E	0
PS38	Haleban	Saudi Arabia	Array	23.4 N	44.5 E	–
PS39	Boshof	South Africa	3C	28.6 S	25.3 E	29
PS40	Sonseca	Spain	Array	39.7 N	4.0 W	0
PS41	Chiang Mai	Thailand	Array	18.5 N	98.9 E	35
PS42	Kesra	Tunisia	3C	35.7 N	9.3 E	55
PS43	Keskin	Turkey	Array	39.7 N	33.6 E	72
PS44	Alibeck	Turkmenistan	Array	37.9 N	58.1 E	0
PS45	Malin	Ukraine	Array	50.7 N	29.2 E	34
PS46	Lajitas, TX	United States of America	Array	29.3 N	103.7 W	0
PS47	Mina, NV	United States of America	Array	38.4 N	118.3 W	9
PS48	Pinedale, WY	United States of America	Array	42.8 N	109.6 W	0
PS49	Eielson, AK	United States of America	Array	64.8 N	146.9 W	0
PS50	Vanda, Antarctica	United States of America	3C	77.5 S	161.9 E	0

* TBD: To be determined

Auxiliary Seismic Stations

Treaty No.	Station	State	Type	LAT	LON	Dev (km)
AS001	Coronel Fontana	Argentina	3C	31.6 S	68.2 W	0
AS002	Ushuaia	Argentina	3C	54.8 S	68.4 W	34
AS003	Garni	Armenia	3C	40.1 N	44.7 E	0
AS004	Charters Towers, QLD	Australia	3C	20.1 S	146.3 E	0
AS005	Fitzroy Crossing, WA	Australia	3C	18.1 S	125.6 E	0
AS006	Narrogin, WA	Australia	3C	32.9 S	117.2 E	0
AS007	Bariadhala, Chittagong	Bangladesh	3C	22.7 N	91.6 E	39
AS008	San Ignacio	Bolivia	3C	16.0 S	61.1 W	0
AS009	Lobatse	Botswana	3C	25.0 S	25.6 E	0
AS010	Pitinga	Brazil	3C	0.7 S	60.0 W	0
AS011	Riachuelo	Brazil	3C	5.8 S	35.9 W	173
AS012	Iqaluit, N.W.T.	Canada	3C	63.7 N	68.5 W	0
AS013	Dease Lake, B.C.	Canada	3C	58.4 N	130.0 W	0
AS014	Sadowa, Ont.	Canada	3C	44.8 N	79.1 W	0
AS015	Bella Bella, B.C.	Canada	3C	52.2 N	128.1 W	0
AS016	Resolute, Nunavut	Canada	3C	74.7 N	94.9 W	699
AS017	Inuvik, N.W.T.	Canada	3C	68.3 N	133.5 W	0
AS018	Easter Island	Chile	3C	27.1 S	109.3 W	15
AS019	Limon Verde	Chile	3C	22.6 S	68.9 W	0
AS020	Baijiatuan	China	3C	40.0 N	116.2 E	0
AS021	Kunming	China	3C	25.1 N	102.7 E	15
AS022	Sheshan	China	3C	31.1 N	121.2 E	0
AS023	Xian	China	3C	34.0 N	108.9 E	0
AS024	Rarotonga	Cook Islands	3C	21.2 S	159.8 W	0
AS025	Las Juntas de Abangares	Costa Rica	3C	10.3 N	85.0 W	0
AS026	Vranov	Czech Republic	3C	49.3 N	16.6 E	0
AS027	Søndre Strømfjord, Greenland	Denmark	3C	67.0 N	50.6 W	0
AS028	Arta Tunnel	Djibouti	3C	11.5 N	42.8 E	11
AS029	Kottamya	Egypt	3C	29.9 N	31.8 E	0
AS030	Furi	Ethiopia	3C	8.9 N	38.7 E	0
AS031	Monasavu, Viti Levu	Fiji	3C	17.7 S	178.1 E	11
AS032	Mont Dzumac	France	3C	22.1 S	166.4 E	10
AS033	Saul, French Guiana	France	3C	3.6 N	53.2 W	187
AS034	Masuku	Gabon	3C	1.7 S	13.6 E	0
AS035	SANAE Station, Antarctica	Germany/South Africa	3C	71.7 S	2.8 W	4
AS036	Anogia, Crete	Greece	3C	35.3 N	24.9 E	0
AS037	El Apazote	Guatemala	3C	15.0 N	90.7 W	22
AS038	Borgames	Iceland	3C	64.7 N	21.3 W	11
AS039	TBD	TBD	TBD			–
AS040	Lembang, Jawa Barat	Indonesia	3C	6.8 S	107.6 E	74
AS041	Jayapura, Irian Jaya	Indonesia	3C	2.5 S	140.7 E	0

(continued)

Treaty No.	Station	State	Type	LAT	LON	Dev (km)
AS042	Sorong, Irian Jaya	Indonesia	3C	0.9 S	131.3 E	0
AS043	Parapat, Sumatera	Indonesia	3C	2.7 N	98.9 E	0
AS044	Kappang, Sulawesi Selatan	Indonesia	3C	5.0 S	119.8 E	0
AS045	Baumata, Timur	Indonesia	3C	10.2 S	123.7 E	11
AS046	Kerman	Iran (Islamic Republic of)	3C	30.0 N	56.8 E	44
AS047	Shushtar	Iran (Islamic Republic of)	3C	32.1 N	48.8 E	52
AS048	Eilath	Israel	3C	29.7 N	35.0 E	15
AS049	Mount Meron	Israel	Array	33.0 N	35.4 E	46
AS050	Valguarnera, Sicily	Italy	3C	37.5 N	14.4 E	9
AS051	Ohita, Kyushu	Japan	3C	33.1 N	130.9 E	0
AS052	Kunigami, Okinawa	Japan	3C	26.8 N	128.3 E	0
AS053	Hachijojima, Izu Islands	Japan	3C	33.1 N	139.8 E	0
AS054	Kamikawa-asahi, Hokkaido	Japan	3C	44.1 N	142.6 E	0
AS055	Chichijima, Ogasawara	Japan	3C	27.1 N	142.2 E	0
AS056	Tel-Alasfar	Jordan	3C	32.2 N	36.9 E	74
AS057	Borovoye	Kazakhstan	Array	53.0 N	70.4 E	13
AS058	Kurchatov	Kazakhstan	Array	50.7 N	78.6 E	0
AS059	Aktyubinsk	Kazakhstan	3C	50.4 N	58.0 E	0
AS060	Ala-Archa	Kyrgyzstan	3C	42.6 N	74.5 E	0
AS061	Ambohidratompo	Madagascar	3C	18.6 S	47.2 E	54
AS062	Kowa	Mali	3C	14.5 N	4.0 W	0
AS063	Tepich, Quintana Roo	Mexico	3C	20.4 N	88.5 W	31
AS064	Colonia Cuauhtémoc Matias Romero, Oaxaca	Mexico	3C	17.1 N	94.9 W	113
AS065	La Paz, Baja California Sur	Mexico	3C	24.1 N	110.3 W	15
AS066	Midelt	Morocco	3C	32.8 N	4.6 W	0
AS067	Tsumeb	Namibia	3C	19.2 S	17.6 E	24
AS068	Everest	Nepal	3C	28.0 N	86.8 E	0
AS069	Rata Peaks, South Island	New Zealand	3C	43.7 S	171.1 E	28
AS070	Raoul Island	New Zealand	3C	29.3 S	177.9 W	11
AS071	Urewera, North Island	New Zealand	3C	38.3 S	177.1 E	0
AS072	Spitsbergen	Norway	Array	78.2 N	16.4 E	0
AS073	Jan Mayen	Norway	3C	71.0 N	8.5 W	13
AS074	Wadi Sarin	Oman	3C	23.2 N	58.6 E	65
AS075	Port Moresby	Papua New Guinea	3C	9.4 S	147.2 E	0
AS076	Keravat	Papua New Guinea	3C	4.3 S	152.0 E	150
AS077	Atahualpa	Peru	3C	7.0 S	78.4 W	44
AS078	Nana	Peru	3C	12.0 S	76.8 W	0

(continued)

Treaty No.	Station	State	Type	LAT	LON	Dev (km)
AS079	Davao, Mindanao	Philippines	3C	7.1 N	125.6 E	0
AS080	Tagaytay, Luzon	Philippines	3C	14.1 N	120.9 E	0
AS081	Muntele Rosu	Romania	3C	45.5 N	25.9 E	0
AS082	Kirov	Russian Federation	3C	58.6 N	49.4 E	0
AS083	Kislovodsk	Russian Federation	Array	44.0 N	42.7 E	0
AS084	Obninsk	Russian Federation	3C	55.1 N	36.6 E	0
AS085	Arti	Russian Federation	3C	56.4 N	58.6 E	0
AS086	Seymchan	Russian Federation	3C	62.9 N	152.4 E	0
AS087	Talaya	Russian Federation	3C	51.7 N	103.6 E	0
AS088	Yakutsk	Russian Federation	3C	62.0 N	129.7 E	0
AS089	Kuldur	Russian Federation	3C	49.2 N	131.8 E	215
AS090	Bilibino	Russian Federation	3C	68.0 N	166.4 E	0
AS091	Tiksi	Russian Federation	3C	71.6 N	128.9 E	0
AS092	Yuzhno-Sakhalinsk	Russian Federation	3C	47.0 N	142.8 E	0
AS093	Magadan	Russian Federation	3C	59.6 N	150.8 E	0
AS094	Zilim	Russian Federation	3C	53.9 N	57.0 E	0
AS095	Afiamalu	Samoa	3C	13.9 S	171.8 W	0
AS096	Dhaban Al-Janub	Saudi Arabia	3C	17.7 N	43.5 E	693
AS097	Babate	Senegal	3C	14.7 N	16.6 W	55
AS098	Honiara, Guadalcanal	Solomon Islands	3C	9.4 S	159.9 E	11
AS099	Sutherland	South Africa	3C	32.4 S	20.8 E	0
AS100	Pallekele	Sri Lanka	3C	7.3 N	80.7 E	99
AS101	Hagfors	Sweden	Array	60.1 N	13.7 E	0
AS102	Davos	Switzerland	3C	46.8 N	9.9 E	8
AS103	Mbarara	Uganda	3C	0.6 S	30.7 E	40
AS104	Eskdalemuir	United Kingdom of Great Britain and Northern Ireland	Array	55.3 N	3.2 W	0
AS105	Guam, Marianas Islands	United States of America	3C	13.6 N	144.9 E	0

(continued)

Treaty No.	Station	State	Type	LAT	LON	Dev (km)
AS106	Palmer Station, Antarctica	United States of America	3C	64.8 S	64.0 W	5
AS107	Tuckaleechee Caverns, TN	United States of America	3C	35.7 N	83.8 W	0
AS108	Piñon Flat, CA	United States of America	3C	33.6 N	116.5 W	0
AS109	Yreka, CA	United States of America	3C	41.7 N	122.7 W	0
AS110	Kodiak Island, AK	United States of America	3C	57.8 N	152.6 W	6
AS111	Albuquerque, NM	United States of America	3C	34.9 N	106.5 W	11
AS112	Attu Island, AK	United States of America	3C	52.9 N	173.2 E	35
AS113	Elko, NV	United States of America	3C	40.7 N	115.2 W	0
AS114	South Pole, Antarctica	United States of America	3C	89.9 S	145.0 W	4
AS115	Newport, WA	United States of America	3C	48.3 N	117.1 W	0
AS116	San Juan, PR	United States of America	3C	18.1 N	66.2 W	0
AS117	Santo Domingo	Venezuela (Bolivarian Republic of)	3C	8.9 N	70.6 W	0
AS118	Puerto la Cruz	Venezuela (Bolivarian Republic of)	3C	10.2 N	64.6 W	0
AS119	Lusaka	Zambia	3C	15.3 S	28.2 E	0
AS120	Matopos	Zimbabwe	3C	20.4 S	28.5 E	–

Radionuclide Stations

Treaty No.	Station	State	Type	LAT	LON	Dev (km)
RN01	Buenos Aires	Argentina	Particulate and Noble Gas	34.5 S	58.5 W	72
RN02	Salta	Argentina	Particulate	24.8 S	65.4 W	98
RN03	Bariloche	Argentina	Particulate	41.1 S	71.2 W	8
RN04	Melbourne, VIC	Australia	Particulate and Noble Gas	37.7 S	145.1 E	49
RN05	Mawson, Antarctica	Australia	Particulate	67.6 S	62.9 E	17
RN06	Townsville, QLD	Australia	Particulate	19.2 S	146.8 E	0
RN07	Macquarie Island	Australia	Particulate	54.5 S	159.0 E	56

(continued)

Treaty No.	Station	State	Type	LAT	LON	Dev (km)
RN08	Cocos Islands	Australia	Particulate	12.2 S	96.8 E	31
RN09	Darwin, NT	Australia	Particulate and Noble Gas	12.4 S	130.9 E	22
RN10	Perth, WA	Australia	Particulate	31.9 S	116.0 E	0
RN11	Rio de Janeiro	Brazil	Particulate and Noble Gas	23.0 S	43.4 W	64
RN12	Recife	Brazil	Particulate	7.8 S	35.1 W	25
RN13	Edea	Cameroon	Particulate and Noble Gas	3.8 N	10.2 E	56
RN14	Sidney	Canada	Particulate	48.7 N	123.5 W	70
RN15	Resolute, NU	Canada	Particulate	74.7 N	95.0 W	3
RN16	Yellowknife, N.W.T.	Canada	Particulate and Noble Gas	62.5 N	114.5 W	0
RN17	St. John's N.L.	Canada	Particulate and Noble Gas	47.6 N	52.7 W	71
RN18	Punta Arenas	Chile	Particulate	53.1 S	70.9 W	20
RN19	Hanga Roa, Easter Island	Chile	Particulate and Noble Gas	27.1 S	109.3 W	89
RN20	Beijing	China	Particulate and Noble Gas	40.0 N	116.4 E	28
RN21	Lanzhou	China	Particulate	36.0 N	104.2 E	84
RN22	Guangzhou	China	Particulate and Noble Gas	23.1 N	113.3 E	11
RN23	Rarotonga	Cook Islands	Particulate	21.2 S	159.8 W	0
RN24	Isla Santa Cruz, Galápagos Islands	Ecuador	Particulate	0.7 S	90.3 W	127
RN25	Addis Ababa	Ethiopia	Particulate and Noble Gas	9.1 N	38.8 E	588
RN26	Nadi	Fiji	Particulate	17.8 S	177.4 E	25
RN27	Papeete, Tahiti	France	Particulate and Noble Gas	17.6 S	149.6 W	79
RN28	Pointe-à-Pitre, Guadeloupe	France	Particulate	16.3 N	61.5 W	95
RN29	Réunion	France	Particulate and Noble Gas	20.9 S	55.6 E	22
RN30	Port-aux-Français, Kerguelen	France	Particulate and Noble Gas	49.4 S	70.3 E	50
RN31	Kourou, French Guiana	France	Particulate and Noble Gas	5.2 N	52.7 W	81
RN32	Dumont dUrville, Antarctica	France	Particulate	66.7 S	140.0 E	78
RN33	Schauinsland/ Freiburg	Germany	Particulate and Noble Gas	47.9 N	7.9 E	0

(continued)

Treaty No.	Station	State	Type	LAT	LON	Dev (km)
RN34	Reykjavik	Iceland	Particulate	64.1 N	21.8 W	34
RN35	TBD	TBD	Particulate			–
RN36	Tehran	Iran (Islamic Republic of)	Particulate and Noble Gas	35.0 N	52.0 E	–
RN37	Okinawa	Japan	Particulate	26.5 N	127.9 E	0
RN38	Takasaki, Gunma	Japan	Particulate and Noble Gas	36.3 N	139.1 E	9
RN39	Kiritimati	Kiribati	Particulate	2.0 N	157.4 W	45
RN40	Kuwait City	Kuwait	Particulate	29.3 N	47.9 E	35
RN41	Misratah	Libyan Arab Jamahiriya	Particulate	32.5 N	15.0 E	–
RN42	Tanah Rata	Malaysia	Particulate	4.5 N	101.4 E	212
RN43	Nouakchott	Mauritania	Particulate and Noble Gas	18.1 N	15.9 W	117
RN44	Guerrero Negro, Baja California	Mexico	Particulate and Noble Gas	28.0 N	114.1 W	108
RN45	Ulaanbaatar	Mongolia	Particulate and Noble Gas	47.9 N	106.3 E	69
RN46	Chatham Island	New Zealand	Particulate and Noble Gas	43.8 S	176.5 W	22
RN47	Kaitaia	New Zealand	Particulate	35.1 S	173.3 E	0
RN48	Agadez	Niger	Particulate and Noble Gas	17.0 N	8.0 E	543
RN49	Spitsbergen	Norway	Particulate and Noble Gas	78.2 N	15.4 E	23
RN50	Panama City	Panama	Particulate and Noble Gas	9.0 N	79.5 W	16
RN51	Kavieng, New Ireland	Papua New Guinea	Particulate	2.6 S	150.8 E	100
RN52	Tanay	Philippines	Particulate	14.6 N	121.4 E	45
RN53	Ponta Delgada, São Miguel, Azores	Portugal	Particulate	37.7 N	25.7 W	43
RN54	Kirov	Russian Federation	Particulate	58.6 N	49.4 E	0
RN55	Norilsk	Russian Federation	Particulate and Noble Gas	69.3 N	87.5 E	39
RN56	Peleduy	Russian Federation	Particulate	59.6 N	112.6 E	0
RN57	Bilibino	Russian Federation	Particulate	68.0 N	166.4 E	0
RN58	Ussuriysk	Russian Federation	Particulate and Noble Gas	44.2 N	132.0 E	56
RN59	Zalesovo	Russian Federation	Particulate	53.9 N	84.8 E	0
RN60	Petropavlovsk-Kamchatskiy	Russian Federation	Particulate and Noble Gas	53.1 N	158.8 E	0

(continued)

Treaty No.	Station	State	Type	LAT	LON	Dev (km)
RN61	Dubna	Russian Federation	Particulate and Noble Gas	56.7 N	37.3 E	0
RN62	Marion Island	South Africa	Particulate and Noble Gas	46.9 S	37.8 E	76
RN63	Stockholm	Sweden	Particulate and Noble Gas	59.4 N	17.9 E	6
RN64	Dar es Salaam	United Republic of Tanzania	Particulate	6.8 S	39.2 E	92
RN65	Bangkok	Thailand	Particulate and Noble Gas	14.0 N	100.0 E	59
RN66	BIOT/Chagos Archipelago	United Kingdom of Great Britain and Northern Ireland	Particulate and Noble Gas	7.3 S	72.4 E	55
RN67	St. Helena	United Kingdom of Great Britain and Northern Ireland	Particulate	15.9 S	5.7 W	34
RN68	Tristan da Cunha	United Kingdom of Great Britain and Northern Ireland	Particulate and Noble Gas	37.1 S	12.3 W	11
RN69	Halley, Antarctica	United Kingdom of Great Britain and Northern Ireland	Particulate and Noble Gas	76.0 S	28.0 W	0
RN70	Sacramento, CA	United States of America	Particulate	38.7 N	121.4 W	0
RN71	Sand Point, Ak	United States of America	Particulate	55.2 N	160.5 W	39
RN72	Melbourne, FL	United States of America	Particulate	28.1 N	80.6 W	22
RN73	Palmer Station	United States of America	Particulate	64.8 S	64.1 W	34
RN74	Ashland, KS	United States of America	Particulate and Noble Gas	37.2 N	99.8 W	0
RN75	Charlottesville, VA	United States of America	Particulate and Noble Gas	38.0 N	78.4 W	35
RN76	Salchaket, AK	United States of America	Particulate	64.7 N	147.1 W	33
RN77	Wake Island	United States of America	Particulate and Noble Gas	19.3 N	166.6 E	0
RN78	Midway Islands	United States of America	Particulate	28.2 N	177.4 W	45

(continued)

Treaty No.	Station	State	Type	LAT	LON	Dev (km)
RN79	Oahu, HI	United States of America	Particulate and Noble Gas	21.5 N	158.0 W	0
RN80	Upi, Guam	United States of America	Particulate	13.6 N	144.9 E	11

Radionuclide Laboratories

Treaty No.	Location	State	Laboratory name
RL01	Buenos Aires	Argentina	National Board of Nuclear Regulation
RL02	Melbourne	Australia	Australian Radiation Protection and Nuclear Safety Agency
RL03	Seibersdorf	Austria	ARC Seibersdorf Research GmBH
RL04	Rio de Janeiro	Brazil	Institute of Radiation Protection and Dosimetry
RL05	Ottawa	Canada	Health Canada
RL06	TBD	China	TBD
RL07	Helsinki	Finland	Radiation and Nuclear Safety Authority
RL08	Bruyères-le-Châtel	France	Atomic Energy Commission
RL09	Yavne	Israel	Soreq Nuclear Research Center
RL10	Rome	Italy	Laboratory of the National Agency for the Protection of the Environment
RL11	Tokai	Japan	Japan Atomic Energy Research Institute
RL12	Christchurch	New Zealand	National Radiation Laboratory
RL13	Moscow	Russian Federation	Central Radiation Control Laboratory
RL14	Pelindaba	South Africa	Atomic Energy Corporation
RL15	Reading	United Kingdom of Great Britain and Northern Ireland	AWE Aldermaston
RL16	Richland	United States of America	Pacific Northwest National Laboratory

Hydroacoustic Stations

Treaty No.	Station	State	Type	LAT	LON	Dev (km)
HA01	Cape Leeuwin, WA	Australia	Hydro-phone	34.3 S	115.2 E	14
HA02	Queen Charlotte Islands, B.C.	Canada	Hydro T-Phase	53.3 N	132.5 W	0

(continued)

Treaty No.	Station	State	Type	LAT	LON	Dev (km)
HA03	Juan Fernández Island	Chile	Hydrophone	33.6 S	78.8 W	11
HA04	Crozet Islands	France	Hydrophone	46.4 S	51.9 E	26
HA05	Guadeloupe	France	Hydro T-Phase	16.3 N	61.1 W	0
HA06	Socorro Island	Mexico	Hydro T-Phase	18.7 N	110.9 W	395
HA07	Flores	Portugal	Hydro T-Phase	39.4 N	31.2 W	14
HA08	BIOT/Chagos Archipelago	United Kingdom of Great Britain and Northern Ireland	Hydrophone	7.3 S	72.4 E	0
HA09	Tristan da Cunha	United Kingdom of Great Britain and Northern Ireland	Hydro T-Phase	37.1 S	12.3 W	21
HA10	Ascension	United Kingdom of Great Britain and Northern Ireland	Hydrophone	8.0 S	14.4 W	0
HA11	Wake Island	United States of America	Hydrophone	19.3 N	166.6 E	0

Infrasound Stations

Treaty No.	Station	State	Type	LAT	LON	Dev (km)
IS01	Villa Traful	Argentina	Array	41.2 S	70.9 W	61
IS02	Ushuaia	Argentina	Array	54.6 S	67.3 W	63
IS03	Davis Base, Antarctica	Australia	Array	68.4 S	77.6 E	0
IS04	Shannon	Australia	Array	34.6 S	116.4 E	203
IS05	Hobart, TAS	Australia	Array	42.5 S	147.7 E	61
IS06	Cocos Islands	Australia	Array	12.2 S	96.8 E	24
IS07	Warramunga, NT	Australia	Array	19.9 S	134.3 E	0
IS08	La Paz	Bolivia	Array	16.2 S	68.5 W	44
IS09	Brasilia	Brazil	Array	15.6 S	48.0 W	0
IS10	Lac du Bonnet, Man.	Canada	Array	50.2 N	96.0 W	7
IS11	Cape Verde Islands	Cape Verde	Array	15.2 N	23.2 W	124
IS12	Bangui	Central African Republic	Array	5.2 N	18.4 E	–
IS13	Easter Island	Chile	Array	27.1 S	109.4 W	23
IS14	Robinson Crusoe Island	Chile	Array	33.6 S	78.8 W	177

(continued)

Treaty No.	Station	State	Type	LAT	LON	Dev (km)
IS15	Beijing	China	Array	39.6 N	115.9 E	45
IS16	Kunming	China	Array	25.3 N	102.7 E	35
IS17	Dimbokro	Côte d'Ivoire	Array	6.7 N	4.9 W	0
IS18	Qaanaaq, Greenland	Denmark	Array	77.5 N	69.3 W	112
IS19	Dijibouti	Djibouti	Array	11.5 N	43.2 E	40
IS20	Isla Santa Cruz, Galápagos Islands	Ecuador	Array	0.6 S	90.4 W	159
IS21	Marquesas Islands	France	Array	8.9 S	140.2 W	125
IS22	Port Laguerre, New Caledonia	France	Array	22.2 S	166.8 E	53
IS23	Kerguelen	France	Array	49.3 S	70.3 E	88
IS24	Tahiti	France	Array	17.8 S	149.3 W	39
IS25	Kourou, French Guiana	France	Array	5.2 N	52.9 W	22
IS26	Freyung	Germany	Array	48.9 N	13.7 E	0
IS27	Georg von Neumayer, Antarctica	Germany	Array	70.7 S	8.3 W	12
IS28	TBD	TBD	Array			–
IS29	Tehran	Iran (Islamic Republic of)	Array	35.7 N	51.4 E	–
IS30	Isumi	Japan	Array	35.3 N	140.3 E	80
IS31	Aktyubinsk	Kazakhstan	Array	50.4 N	58.0 E	0
IS32	Nairobi	Kenya	Array	1.3 S	36.8 E	0
IS33	Antananarivo	Madagascar	Array	19.0 S	47.3 E	31
IS34	Songino	Mongolia	Array	47.8 N	106.4 E	37
IS35	Tsumeb	Namibia	Array	19.2 S	17.6 E	24
IS36	Chatham Island	New Zealand	Array	43.9 S	176.5 W	11
IS37	Karasjok	Norway	Array	69.5 N	25.5 E	–
IS38	Rahimyar Khan	Pakistan	Array	28.2 N	70.3 E	–
IS39	Palau	Palau	Array	7.5 N	134.5 E	0
IS40	Keravat	Papua New Guinea	Array	4.3 S	152.0 E	25
IS41	Villa Florida	Paraguay	Array	26.3 S	57.3 W	0
IS42	Azores, Graciosa	Portugal	Array	39.0 N	28.0 W	256
IS43	Dubna	Russian Federation	Array	56.7 N	37.3 E	0
IS44	Petropavlovsk-Kamchatskiy	Russian Federation	Array	53.1 N	157.7 E	74
IS45	Ussuriysk	Russian Federation	Array	44.2 N	132.0 E	56
IS46	Zalesovo	Russian Federation	Array	53.9 N	84.8 E	0
IS47	Boshof	South Africa	Array	28.6 S	25.3 E	10

(continued)

Treaty No.	Station	State	Type	LAT	LON	Dev (km)
IS48	Kesra	Tunisia	Array	35.8 N	9.3 E	59
IS49	Tristan da Cunha	United Kingdom of Great Britain and Northern Ireland	Array	37.1 S	12.3 W	11
IS50	Ascension	United Kingdom of Great Britain and Northern Ireland	Array	7.9 S	14.4 W	16
IS51	Bermuda	United Kingdom of Great Britain and Northern Ireland	Array	32.3 N	64.7 W	38
IS52	BIOT/Chagos Archipelago	United Kingdom of Great Britain and Northern Ireland	Array	7.4 S	72.5 E	273
IS53	Fairbanks, AK	United States of America	Array	64.9 N	147.9 W	49
IS54	Palmer Station, Antarctica	United States of America	Array	64.8 S	64.1 W	1386
IS55	Windless Bight, Antarctica	United States of America	Array	77.7 S	167.6 E	140
IS56	Newport, WA	United States of America	Array	48.3 N	117.1 W	0
IS57	Piñon Flat, CA	United States of America	Array	33.6 N	116.5 W	0
IS58	Midway Islands	United States of America	Array	28.2 N	177.4 W	23
IS59	Hawaii, HI	United States of America	Array	19.6 N	155.9 W	63
IS60	Wake Island	United States of America	Array	19.3 N	166.6 E	–

Annex 2

This annex lists Chairpersons and Executive Secretaries of the CTBTO Preparatory Commission, Chairpersons, Focal Points and Friends of the Chair of Working Group A, as well as Chairpersons, Friends of the Chair, Program Coordinators and Task Leaders of Working Group B of the CTBTO Preparatory Commission from 1997 to date. The Chairpersons have been elected by the CTBTO Preparatory Commission. Focal Points, Friends of the Chair, Program Coordinators and Task Leaders have been named by the Chairperson of the respective Working Groups and are or have been exercising their functions on behalf of the Chairperson.

Chairpersons of the CTBTO Preparatory Commission

Task	Name	State signatory	Period
Chairing the CTBTO Preparatory Commission on behalf of the States Signatories	Amb. Jacob S. Selebi	South Africa	Nov 1996–May 1997
	Amb. Daniela Rozgonova	Slovakia	May–Dec 1997
	Amb. Affonso Celso de Ouro-Preto	Brazil	Jan–Jun 1998
	Amb. John P. G. Freeman	United Kingdom	Jul–Dec 1998
	Amb. Ban Ki-moon	Republic of Korea	Jan–Jun 1999
	Amb. Mokhtar Reguieg	Algeria	Jul–Dec 1999
	Amb. Pavel Vacek	Czech Republic	Jan–Jun 2000
	Amb. Olga Pellicer	Mexico	Jul–Dec 2000
	Amb. Jaap Ramaker	The Netherlands	Jan–Jun 2001
	Amb. R. I. Rhousdy Soeriaatmadja	Indonesia	Jul–Dec 2001
	Amb. Abdul Bin Rimdap	Nigeria	Jan–Jun 2002

	(continued)		
Task	Name	State signatory	Period
	Amb. Liviu Aurelian Bota	Romania	Jul–Dec 2002
	Amb. Javier Paulinich	Peru	Jan–Jun 2003
	Amb. Thomas Stelzer	Austria	Jul–Dec 2003
	Amb. Yukio Takasu	Japan	2004
	Amb. Taous Feroukhi	Algeria	2005
	Amb. Volodymyr Yelchenko	Ukraine	2006
	Amb. Ana Teresa Dengo	Costa Rica	2007
	Amb. Hans Lundborg	Sweden	2008

Executive Secretaries of the CTBTO Preparatory Commission

Task	Name	State signatory	Period
Executive Secretary of the CTBTO Preparatory Commission	Amb. Wolfgang Hoffmann	Germany	Mar 1997–Jul 2005
	Amb. Tibor Tóth	Hungary	Aug 2005...

Chairpersons of Working Group A

Task	Name	State signatory	Period
Chairing Working Group A on behalf of the States Signatories	Amb. Tibor Tóth	Hungary	1997–2005
	Amb. Patricia Espinosa Cantellano	Mexico	2005
	Amb. Abdul Bin Rimdap	Nigeria	2005...

Focal Points and Friends of the Chair of Working Group A

Task	Name	State signatory	Period
Focal Point on Advisory Group and related issues	Mr André Gué	France	1997
Focal Point on reduced assessment	Mr Donald Phillips	United States of America	1997
Focal Point/Friend of the Chair on model arrangements, taxation, privileges and immunities for the PrepCom and its	Mr Trevor Moore	United Kingdom	1997–2001

(continued)

Task	Name	State signatory	Period
officials, relationship agreement between the CTBTO PrepCom and the UN, post-certification costs of certain new auxiliary seismic stations and legal basis for post-certification costs			
Friend of the Chair on late payments	Ms Lucy Duncan	New Zealand	1998–1999
Friend of the Chair on legal procedures for dealing with alternate site locations and station names/codes	Mr Charles Oleszycki	United States of America	1998–2002
Friend of the Chair on post-certification costs of certain new auxiliary seismic stations	Ms Susan Coles	Australia	2001
Focal point on possible reduction of the annual number of sessions of the PrepCom and its subsidiary bodies from 2003 onwards	Mr Mohammad Daryaei	Iran (Islamic Republic of)	2002
Focal Point on the follow-up response of the PTS to the external evaluation of human resources issues in the PTS	Ms Tracy Roberts	United Kingdom	2002
Focal Point on status of on-site inspection inspectors and inspection assistants	Mr Bernard Bourges	France	2002–2004
	Mr Gustavo Gonzalez	Chile	2002
	Ms Maria Feliciana Ortigão	Brazil	2004–2006

Chairpersons of Working Group B

Task	Name	State signatory	Period
Chairing Working Group B on behalf of the States Signatories	Mr Ola Dahlman	Sweden	1997–2006
	Mr Hein Haak	The Netherlands	2006...

Friends of the Chair of Working Group B

Task	Name	State signatory	Period
Advising and assisting the Chairperson in the conduct of Working Group B business	Mr Hein Haak	The Netherlands	1997–2006
	Mr Svein Mykkeltveit	Norway	1997...
	Mr David A. McCormack	Canada	2006...

Program Coordinators of Working Group B

Major program	Name	State signatory	Period
International Monitoring System	Mr Shigeji Suyehiro	Japan	1997
	Mr John Alwyn Davies	United Kingdom	1997–2000
	Mr Ralph W. Alewine III	United States of America	2000–2002
	Mr Jay Zucca	United States of America	2003...
International Data Center	Mr Ralph W. Alewine III	United States of America	1997–2000
	Mr Ken Muirhead	Australia	2000–2001
	Mr Frode Ringdal	Norway	2001...
Communications	Mr Manfred Henger	Germany	1997
	Mr Artur Schechtman	Brazil	1997
	Mr Hans Frese	Germany	1998...
On-Site Inspection	Mr Vitaliy N. Shchukin	Russian Federation	1997...
Evaluation	Mr Bernard Massinon	France	1998–1999
	Ms Marilyn White	Canada	1999–2002
	Mr Hakim Gheddou	Algeria	2003
	Mr David A. McCormack	Canada	2004–2006
	Mr Hans Frese	Germany	2007...

Task Leaders of Working Group B

Task	Name	State signatory	Period
GSETT-3 and Other Ongoing Tests	Mr Robert Kleywegt	South Africa	1997
Quality Assurance	Mr Dominik Pacala	Slovakia	1997
Event Screening Criteria	Mr Ralph W. Alewine III	United States of America	1997
	Mr Wang Hong	China	1997
IMS Training Program	Mr Mohsen Ghafory-Ashtiany	Iran (Islamic Republic of)	1997
Plans for Certifying Stations	Mr Yves Caristan	France	1997

(continued)

Task	Name	State signatory	Period
Station Configuration for Infrasound and Hydroacoustic Stations	Mr Ken Muirhead	Australia	1997
Procedures for Cooperating National Facilities	Mr Peter Basham	Canada	1997
IMS Facilities Inventory	Mr Ken Muirhead	Australia	1997
Authentication	Mr Yves Caristan	France	1997–1998
	Mr Bernard Massinon	France	1999–2000
Model Agreements or Arrangements	Mr Ken Muirhead	Australia	1997–1999
Draft IMS Operational Manuals	Mr Peter Basham	Canada	1997
	Mr Malcolm Cooper	Australia	1998–2000
	Ms Victoria Oancea	Romania	1997...
Draft IDC Operational Manual	Mr Ralph W. Alewine III	United States of America	1997
	Mr David A. McCormack	Canada	2000–2006
	Ms Victoria Oancea	Romania	2007...
Draft OSI Operational Manual	Mr Vitaliy Shchukin	Russian Federation	1997
	Mr Arend J. Meerburg	The Netherlands	2001–2004
	Mr Malcolm Coxhead	Australia	2004...
Noble Gas Stations	Mr Antonio Oliveira	Argentina	1997–1998
Confidence Building Measures	Mr Ken Muirhead	Australia	1998–1999
Refinements to Station Coordinates	Mr Ken Muirhead	Australia	1998–2000
Radionuclide Issues/ Radionuclide Laboratories	Mr Antonio Oliveira	Argentina	1998–2004
	Mr Carlos Nollmann	Argentina	2001–2004
Science and Technology	Mr Mohsen Ghafory-Ashtiany	Iran (Islamic Republic of)	1998–2007
Reduced Assessments	Ms Lucy Duncan	New Zealand	1998–1999
	Mr Ryotaro Suzuki	Japan	2000–2001
	Mr Hideki Ishizuka	Japan	2001–2003
	Mr Kazuto Suda	Japan	2003–2007
	Mr Takeshi Koizumi	Japan	2007...
Confidentiality	Mr René Haug	Switzerland	1998–1999
	Mr Bernard Massinon	France	1999–2004
Testing and Certification	Mr Ken Muirhead	Australia	1999
Access to IMS Data and IDC Products	Mr Bernard Massinon	France	2000–2004

(continued)

Task	Name	State signatory	Period
Consultation and Clarification	Mr John Alwyn Davies	United Kingdom	2000–2004
Data and Products for Emergency Response	Mr Bernard Massinon	France	2005...
Technology Refreshment and Recapitalization	Mr Jay Zucca	United States of America	2005...
Performance Assessment	Mr David A. McCormack	Canada	2005–2006
	Mr Hans Frese	Germany	2007...
Testing and Provisional Operation	Mr Frode Ringdal	Norway	2005...
Issues Related to NDCs	Mr Mohsen Ghafory-Ashtiany	Iran (Islamic Republic of)	2005–2007
	Mr Frode Ringdal	Norway	2007
	Mr Norbert Opiyo-Akech	Kenya	2007...
Common Issues Related to IFE	Mr Vitaliy N. Shchukin	Russian Federation	2005...

Abbreviations and Acronyms

ABM	Anti-Ballistic Missile
ACS	Association of Caribbean States
AFTAC	Air Force Technical Applications Center (USA)
AG	Advisory Group (of the CTBTO Preparatory Commission)
ARPA	Advanced Research Projects Agency (USA)
BWC	Biological Weapons Convention
CBM	Confidence Building Measures
CCD	Conference of the Committee on Disarmament
CD	Conference on Disarmament
CIF	Capital Investment Fund (of the CTBTO Preparatory Commission)
CMR	Center for Monitoring Research (USA)
CNF	Cooperating National Facility
CTBT	Comprehensive Nuclear-Test-Ban Treaty
CTBTO	Comprehensive Nuclear-Test-Ban Treaty Organization
CWC	Chemical Weapons Convention
DARPA	Defense Advanced Research Projects Agency (USA)
DG	Director General
EC	Executive Council, of the CTBTO
ECMWF	European Centre for Medium-Range Weather Forecasts
ECS	Experts Communication System (of the CTBTO Preparatory Commission)
EIDC	Experimental International Data Center
EIF	Entry into Force (of the CTBT)
EMP	Electro-Magnetic Pulse
EMSC	European-Mediterranean Seismological Centre
ENCD	United Nations Eighteen Nation Committee on Disarmament
ES	Executive Secretary (of the CTBTO Preparatory Commission)
EU	European Union
FMCT	Fissile Material Cut-Off Treaty
FOI/FOA	Swedish Defence Research Agency
G77	Group of 77 (and China)
GCI	Global Communications Infrastructure
GPS	Global Positioning System
GSE	Group of Scientific Experts
GSETT-1	Group of Scientific Experts Technical Test No. 1
GSETT-2	Group of Scientific Experts Technical Test No. 2

GSETT-3	Group of Scientific Experts Technical Test No. 3
GTS	Global Telecommunication System
IAEA	International Atomic Energy Agency
ICBM	Intercontinental Ballistic Missile
IDC	International Data Centre
IFE	Integrated Field Exercise
ILO	International Labour Organization
IMS	International Monitoring System
INGE	International Noble Gas Experiment
IPCC	Intergovernmental Panel on Climate Change
IRIS	Incorporated Research Institutions for Seismology
ISC	International Seismological Centre
JMA	Japan Meteorological Agency
KNMI	Royal Netherlands Meteorological Institute
kt	Kiloton of TNT
MAD	Mutual Assured Destruction
NAS	National Academy of Sciences (USA)
NASA	National Aeronautics and Space Agency (USA)
NCEP	National Centers for Environmental Prediction (USA)
NDC	National Data Center
NEIC	National Earthquake Information Center (USA)
NOAA	National Oceanic and Atmospheric Administration (USA)
NORSAR	Originally short for Norwegian Seismic Array, now the name of a research institute
NPT	Non-Proliferation Treaty
NTS	Nevada Test Site (USA)
OAU	Organization of African Unity
OPANAL	Agency for the Prohibition of Nuclear Weapons in Latin America and the Caribbean
OPBW	Organisation for the Prohibition of Biological Weapons
OPCW	Organisation for the Prohibition of Chemical Weapons
OSI	On-Site Inspection
PIDC	Prototype International Data Center
PNE	Peaceful Nuclear Explosion
PNET	Peaceful Nuclear Explosion Treaty
PrepCom	Preparatory Commission for the CTBTO
PTBT	Partial (or Limited) Test Ban Treaty
PTS	Provisional Technical Secretariat (of the CTBTO Preparatory Commission)
REB	Reviewed Event Bulletin
RRR	Reviewed Radionuclide Report
RRW	Reliable Replacement Warhead
S&T	Science and Technology
SALT	Strategic Arms Limitation Talks
SNR	Signal-to-Noise Ratio
SOFAR	Sound Fixing and Ranging
SOSUS	Sound Surveillance System
SPT1	System-Wide Performance Test 1
TNT	Trinitrotoluene explosive

TS	Technical Secretariat
TTBT	Threshold Test Ban Treaty
UNDP	United Nations Development Programme
UNGA	United Nations General Assembly
UNIDO	United Nations Industrial Development Organisation
UNMOVIC	United Nations Monitoring Verification and Inspection Commission
USGS	U.S. Geological Survey
USSR	Union of Socialistic Soviet Republics
Vela	Name of US verification program
VIC	Vienna International Centre
WGA	Working Group A, of the CTBTO Preparatory Commission
WGB	Working Group B, of the CTBTO Preparatory Commission
WMO	World Meteorological Organization
WWSSN	World Wide Standard Station Network

References*

A/RES/50/64 (1995) UN General Assembly

A749/27 (1994) Official Records of the UN General Assembly

ACA (2007) US Arms Control Association, Nuclear Weapons: Who Has What at a Glance. http://www.armscontrol.org/factsheets/Nuclearweaponswhohaswhat.asp

Adushkin VV, Khristoforov BD (1995) About the Control of the Underwater and Above Water Nuclear Explosions by Hydroacoustic Methods, Russian Academy of Sciences, Moscow Institute for Dynamics of Geospheres. http://stinet.dtic.mil/oai/oai? &verb = getRecord&metadataPrefix = html&identifier = ADP204516

Aki K, Richards PG (2002) Quantitative Seismology. 2nd ed. University Science Books

Atomic Archive (2007a) http://www.atomicarchive.com/Docs/Deterrence/BaruchPlan.shtml

Atomic Archive (2007b) http://www.atomicarchive.com/Treaties/Treaty7.shtml

Avenhaus R, Kremenyuk VA, Sjöstedt G (2002) Containing the Atom. Lexington Books

Bahavar M, Bowman JR, Israelsson HG, Kohl BC, North RG, O'Brian MS, Shields G (2006) Infrasound resources of the SMDC monitoring research program. In: Proceedings of the 28th Seismic Research Review: Ground-based nuclear explosion monitoring technologies. LA-UR-06-5471, Vol. 2, pp. 863–872. https://www.nemre.nnsa.doe.gov/prod/researchreview/2006/PAPERS/06-01.PDF

Bannister RW, Guthrie KM, Kay JS, Bold GES, Johns MD, Tan SM, Tindle CT (1993) ATOC- New Zealand receiver site survey and acoustic test. JASA 93(4):2380

Barker B, Clark M, Davis P, Fisk M, Hedlin M, Israelsson H, Khalturin V, Kim W-Y, McLaughlin K, Meade C, Murphy J, North R, Orcutt J, Powell C, Richards PG, Stead R, Stevens J, Vernon F, Wallace T (1998) Monitoring nuclear tests. Science 281:1967–1968

Barth K-H (2003) The Politics of Seismology. Nuclear Testing, Arms Control, and the Transformation of a Discipline. Social Studies of Science 33:743–781

BBC (2007) http://news.bbc.co.uk/1/hi/world/asia-pacific/2340405.stm

Becker A, Wotawa G, De Geer L-E, Seibert P, Draxler R, Sloan C, D'Amours R, Hort M, Glaab H, Heinrich P, Grillon Y, Shershakov V, Katayama K, Zhang Y, Stewart P, Hirtl M, Jean M,. Chen P (2007) Global backtracking of anthropogenic radionuclides by means of a receptor oriented ensemble dispersion modelling system in support of Nuclear-Test-Ban Treaty verification. Atmospheric Environment 41:4520

Beckett M (2007) Keynote Address; A World Free of Nuclear Weapons? Carnegie Endowment for International Peace. 25 June http://www.carnegieendowment.org/events/index.cfm?fa = eventDetail&id = 1004&&prog = zgp&proj = znpp

Bidwai P, Vanaik A (2000) New Nukes: India, Pakistan and Global Disarmament. Signal Books

Bolt BA (1999) Earthquakes. Freeman

British Antarctic Survey (2007) http://www.antarctica.ac.uk/About_Antarctica/Treaty/

---

* The hyperlinks were accessed in January 2008

Brooks LF (2005) Brooks statement to Senate Armed Services Committee, April 4 http://www.nnsa.doe.gov/docs/congressional/2005/2005-04-04_Brooks_SASC_testimony.pdf

CCD/558 (1978) First Report of the Ad Hoc Group of Scientific Experts to Consider International Co-operative Measures to Detect and Identify Seismic Events. Conference of the Committee on Disarmament. 7 March, Geneva

CD/1089 (1991) Letter dated 91/07/09 from the head of the Swedish Delegation addressed to the Secretary-General of the Conference on Disarmament transmitting the text of a draft Comprehensive Test-Ban Treaty and its annexed protocols. Conference on Disarmament, Geneva

CD/1144 (1992) Report on the Group of Scientific Experts' Second Technical Test (GSETT-2): 6th report to the Conference on Disarmament of the Ad Hoc Group of Scientific Experts to Consider International Cooperative measures to detect and identify Seismic Events. 13 March, Geneva

CD/1212 (1993) Decision on agenda item 1 "Nuclear Test Ban" / adopted by the Conference on Disarmament at its 659th plenary meeting on 10 August 1993. Conference on Disarmament, Geneva

CD/1235 (1994) Letter dated 94/01/04 from the Permanent Representative of Australia to the United Nations for disarmament matters addressed to the President of the Conference on Disarmament transmitting the text of a working paper entitled "Comprehensive nuclear test ban". Conference on Disarmament, Geneva

CD/1238 (1994) Mandate for an Ad Hoc Committee under agenda item 1: "Nuclear test ban". Conference on Disarmament, Geneva

CD/1254 (1994) Report of the Ad Hoc Group of Scientific Experts to Consider International Cooperative Measures to Detect and to Identify Seismic Events to the Ad Hoc Committee on Nuclear Test Ban on International Seismic Monitoring and the GSETT-3 Experiment. 25 March, Geneva

CD/1364 (1995) Report of the Conference on Disarmament to the General Assembly of the United Nations. Conference on Disarmament. 26 September, Geneva

CD/1372 (1995) Progress Report to the Conference on Disarmament on the Forty-Second Session of the Ad Hoc Group of Scientific Experts to Consider International Measures to Detect and Identify Seismic Events. 21 December, Geneva

CD/1384 (1996) Draft comprehensive nuclear test ban treaty: Islamic Republic of Iran. Conference on Disarmament, Geneva

CD/1386 (1996) Comprehensive nuclear test ban treaty: model treaty text: Australia. Conference on Disarmament, Geneva

CD/1423 (1996) Report of the Ad Hoc Group of Scientific Experts to the Conference on Disarmament on the GSETT-3 experiment and its relevance to the seismic component of the comprehensive nuclear test-ban treaty international monitoring system 15 August, Geneva

CD/1427 (1996) Letter dated 96/08/22 from the Permanent Representative of Belgium addressed to the President of the Conference transmitting the text of a draft Comprehensive Nuclear-Test-Ban Treaty. Conference on Disarmament, Geneva

CD/43 (1979) Letter dated 25 July 1979 from the Chairman of the Ad Hoc Group of Scientific Experts to Consider International Co-operative Measures to Detect and Identify Seismic Events to the Chairman of the Committee on Disarmament transmitting the Second Report of the Ad Hoc Group. 25 July, Geneva

CD/448 (1984) Third report to the Conference on Disarmament of the Ad Hoc Group of Scientific Experts to Consider International Co-operative Measures to Detect and to Identify Seismic Events 9 March, Geneva

CD/756 (1987) Letter dated 87/06/08 from the representatives of Bulgaria, Czechoslovakia, the German Democratic Republic, Hungary, Mongolia, Poland, Romania and the Union of Soviet Socialist Republics, addressed to the President of the Conference on Disarmament, transmitting the text of the "Basic provisions of a treaty on the complete and general prohibition of nuclear weapons tests". Conference on Disarmament, Geneva

CD/903 (1989) Fifth report to the Conference on Disarmament of the Ad Hoc Group of Scientific Experts to Consider International Co-operative Measures to Detect and to Identify Seismic Events: technical concepts for a global system for international seismic data exchange. 17 March, Geneva

CD/NTB/WP.224 (1995) International Monitoring System, Expert Group Report based on Technical Discussions held from 6 February to 3 March 1995. Conference on Disarmament, Geneva

CD/NTB/WP.269 (1995) International Monitoring System, Expert Group Report based on Technical Discussions held from 22 to 25 August 1995. Conference on Disarmament, Geneva

CD/NTB/WP.283 (1995) International Monitoring System, Report of the Expert Group based on Technical Discussions held from 4 through 15 December 1995. Conference on Disarmament, Geneva

CD/NTB/WP.330 (1996) Draft Comprehensive Nuclear Test-Ban Treaty: working paper/ Chairman of the Ad Hoc Committee on a Nuclear Test Ban. Conference on Disarmament, Geneva

CD/NTB/WP.330/Rev.2 (1996) Draft Comprehensive Nuclear Test-Ban Treaty / Chairman of the Ad Hoc Committee on a Nuclear Test Ban. Conference on Disarmament, Geneva

Clancy T (1984) The Hunt for Red October. US Naval Institute Press

CNS (2007a) http://cns.miis.edu/pubs/inven/pdfs/spnfz.pdf

CNS (2007b) http://cns.miis.edu/pubs/inven/pdfs/seanwfz.pdf

CNS (2007c) http://cns.miis.edu/pubs/inven/pdfs/aptanwfz.pdf

CNS (2007d) http://cns.miis.edu/pubs/inven/pdfs/atosuw.pdf

Condon EU (1943) Los Alamos Primer, available at: http://www.cfo.doe.gov/me70/manhattan/publications/LANLSerberPrimer.pdf

Convention on the Territorial Sea and Contiguous Zone (1958). http://untreaty.un.org/ilc/texts/instruments/english/conventions/8_1_1958_territorial_sea.pdf

Cox C (1999) PRC Theft of U.S. Thermonuclear Warhead Design Information, Ch. 2, available at http://www.house.gov/coxreport/

CTBT (1996) Comprehensive Nuclear-Test-Ban Treaty. http://www.ctbto.org/treaty/treaty_text.pdf

CTBT/PC-24/INF.9 (2005) Final report on the review of the organizational structure of the Provisional Technical Secretariat of the CTBTO Preparatory Commission. 22 April

CTBT/PTS/INF.872 (2007) CTBTO Medium Term Plan 2008–2012

CTBT/WGB-14/INF.3 (2000) Report of the External Evaluation Team on the International Data Center. 24 November

CTBT/WGB-17/INF.3 (2001) Report of the External Evaluation Team on the International Monitoring System. 21 December

CTBT/WGB-21/INF.5 (2003) Report on the External Evaluation of Major Programme 4: On-Site Inspection. 5 June

CTBTO Preparatory Commission (2005) Regulations and Rules of the Preparatory Commission for the Comprehensive Nuclear-Test-Ban Treaty Organization, Second Edition June

CTBTO Preparatory Commission (2006) CTBT: Synergies with Science 1996–2006 and beyond. ISBN: 92-95021-34-7, CTBTO Preparatory Commission

CTBTO Preparatory Commission (2007a) http://www.ctbto.org/reference/legal_resources/prepcom_resolution.pdf

CTBTO Preparatory Commission (2007b) Potential civil and scientific applications of the CTBT verification technologies. http://www.ctbto.org/reference/outreach/civil_and_scientific_a_2007_afc_web.pdf

CTBTO UNGA (2000) Agreement to regulate the relationship between the United Nations and the Preparatory Commission of the CTBTO. UN General Assembly A/RES/54/280 30 June

D'Agostino TP (2006) Statement by Thomass P. D'Agostino, deputy Administrator for Defense, Programs National Nuclear Security Administration before the House Armed

Services Committee, Subcommittee on Strategic Forces, April 5, http://www.nnsa.doe.gov/docs/congressional/2006/2006-04-05_HASC_Transformation_Hearing_Statement_(DAgostino).pdf

Dahlen FA, Tromp J (1998) Theoretical Global Seismology. Princeton University Press

Dahlman O, Israelsson H (1977) Monitoring Underground Nuclear Explosions. Elsevier

De Geer L-E (1991) Observations in Sweden of Venting Underground Nuclear Explosions. Proceedings from the Symposium on Underground Nuclear Weapons Testing: Potential Environmental Impacts and Their Containment. Ottawa, Canada 23–24 April

De Geer L-E (2001) Comprehensive Nuclear-Test-Ban Treaty: relevant radionuclides. Kerntechnik 66(3):113–120

Decode Systems (2007) http://www.decodesystems.com/gps.html

Dewey J, Byerly P (2007) The early history of seismometry (to 1900) http://earthquake.usgs.gov/learning/topics/seismology/history/part13.php

DOE (2000) United States Nuclear Tests. July 1945 through September 1992. DOE/NV-209 (Rev. 15), December. http://www.fas.org/resource/08132004140930.pdf

DOE (2007) Plowshare Program. US Department of Energy, Nevada Operations Office. https://www.osti.gov/opennet/reports/plowshar.pdf

DOS (1999) The Stockpile Stewardship Program, U.S. Department of State http://www.state.gov/www/global/arms/factsheets/wmd/nuclear/ctbt/fs_991008_stockpile.html

Douglas A, Bowers D, Marshall PD, Young JB, Porter D, Wallis NJ (1999) Putting nuclear-test monitoring to the test. Nature 398:474–475

Ebeling C, Astiz L, Starovoit Y, Tavener N, Perez G, Given HK, Barrientos S, Yamamoto M, Hfaiedh M, Stewart R, Estabrook C (2002) Low noise results from IMS site surveys: A preliminary new high-frequency low noise model. EOS – Transactions of the American Geophysical Union 83(47), Fall Meeting Supplement, Abstract #S71A-1057

EMSC (2007) http://www.emsc-csem.org

Estabrooks S (2006) Getting back to work? The P6 initiative and informal debates in the Conference on Disarmament. The Ploughshares Monitor. Summer 2006, 27(2) http://www.ploughshares.ca/libraries/monitor/monj06e.pdf

FAS (1990) Treaty Between the United States of America and the Union of Soviet Socialist Republics on the Limitation of Underground Nuclear Weapon Tests. http://www.fas.org/nuke/control/ttbt/text/ttbt1.htm

FAS (2007) Federation of American Scientists. Nuclear Weapon Design http://www.fas.org/nuke/intro/nuke/design.htm

FAS abm (2007) http://www.fas.org/nuke/control/abmt/text/abm2.htm

FAS salt 1 (2007) http://www.fas.org/nuke/control/salt1/text/salt1.htm

FAS ttbt (2007) http://www.fas.org/nuke/control/ttbt/text/ttbt1.htm

Feffer J (2007) North Korean Nuclear Agreement: Annotated Foreign Policy in Focus. http://www.fpif.org/fpiftxt/3997

FOI (2006) Press release dated 19 December entitled 'FOI found radioactive xenon following explosion in North Korea'. Swedish Defence Research Agency http://www.foi.se/FOI/Templates/NewsPage____5412.aspx

Foster JS (2003) Panel FY 2003, Report to Congress of the Panel to Assess the Reliability, Safety, and Security of the United States Nuclear Stockpile. April 11 http://www.fas.org/main/content.jsp?formAction = 297&contentId = 256

Gitterman Y, Shapira A (2001) Dead Sea seismic calibration experiment contributes to CTBT monitoring. Seismological Research Letters 72:159–170

Glasstone S, Dolan PJ (1983) The Effects of Nuclear Weapons. United States Government Printing

Global Security (2005) Reliable Replacement Warhead http://www.globalsecurity.org/wmd/systems/rrw.htm

Global Security (2007) http://www.globalsecurity.org/wmd/intro/hydrodynamic.htm

Goldblat J (1988) Nuclear Weapon Tests: Prohibition or limitation? Oxford University Press

Goldblat J (2000) Nuclear Disarmament: Obstacle to Banishing the Bomb. I.B. Taunis

Goldblat J (2002) Arms Control. The new Guide to Negotiations and Agreements. Sage Publications Inc

Goldman J, Stein G (1997) The Cuban Missile crisis October 18–29, 1962 http://www.hpol.org/jfk/cuban/

Gossard EE, Hooke WH (1975) Waves in the Atmosphere: Atmospheric Infrasound and Gravity waves- their Generation and Propagation. ISBN 0-444-411968 Elsevier

Green PE, Frosch RA, Romney CF (1965) Principles of an experimental large aperture seismic array (LASA). Proceedings of the IEEE. Vol. 53/12 pp. 1821–1833

Gutenberg B (1946) Interpretation of Records obtained from the New Mexico Atomic Bomb Test, July 16, 1945. Bulletin of the Seismological Society of America 36:327–330

Hafemeister D (2007) Progress in CTBT monitoring since its 1999 Senate defeat. Science and Global Security 15:151–183

Hall LH (2007) http://www.ctbtcommission.org/hallpaper.htm

Hansen KA (2006) The Comprehensive Nuclear Test Ban Treaty. An Insider's Perspective. Stanford University Press, Stanford, California

Heeger D (1997) Signal Detection Theory. http://www.cns.nyu.edu/~david/sdt/sdt.html

Hobson D (2004) Remarks by Chairman David Hobson House Appropriations Subcommittee on Energy and Water Development, to the National Academy of Sciences Symposium on Post-Cold War US Nuclear Strategy: A Search for Technical and Policy common Ground August 11 http://www7.nationalacademies.org/cisac/

IAEA (1970) http://www.iaea.org/Publications/Documents/Infcircs/Others/infcirc140.pdf

IAEA (2007a) http://www.iaea.org/OurWork/SV/Safeguards/sg_protocol.html

IAEA (2007b) http://www.iaea.org/Publications/Documents/Treaties//tlatelolco.html

ICBL (2007) http://www.icbl.org/

India Nuclear Update (2003) http://www.wisconsinproject.org/countries/india/nuke2003.htm

Inventors (2007) http://inventors.about.com/library/inventors/blseismograph2.htm

IPCC (2007) http://www.ipcc.ch/about/index.htm

IRIS (2007) http://www.iris.washington.edu

ISC (2007) http://www.isc.ac.uk

ISIS (2001) http://www.isis-online.org/publications/southafrica/03012001%20press%20release%20on%20flash.html

Issartel J-P, Baverel J (2003) Inverse transport for the verification of the Comprehensive Nuclear-Test-Ban Treaty. Atmospheric Chemistry and Physics 3:475–486. http://www.atmos-chem-phys.org/aco/3/475/

Jans W (2005) http://www.vertic.org/assets/WEBSITE%20-%20CTBT%20Seminar/Wolfgang%20Jans%20-%2022%20Sept%202005%20-%20NY%20CTBT.pdf

Jeanloz R (2000) Science-Based Stockpile Stewardship. Physics Today 53(12):44–50. http://www.physicstoday.org/pt/vol-53/iss-12/p44.html

Jeanloz R (2007) Comprehensive Nuclear Test Ban Treaty and U.S. Security. http://media.hoover.org/documents/Drell_Goodby_Schultz_Reykjavik_Revisited_55.pdf

JMA (2007) http://www.jma.go.jp/en/quake

Kim W-Y, Richards PG (2007) North Korean Nuclear Test: Seismic discrimination at low yield. EOS – Transactions of the American Geophysical Union 88(14):158, 161

Kimball D, Boese W (2003) Limited Test Ban Treaty turns 40 http://www.armscontrol.org/act/2003_10/LTBT.asp

Kværna T, Ringdal F (1999) Seismic threshold monitoring for continuous assessment of global detection capability. Bulletin of the Seismological Society of America 89:946–959

Kværna T, Ringdal F, Baadshaug U (2007) North Korea's nuclear test: The capability for seismic monitoring of the North Korean test site. Seismological Research Letters 78:487–497

Lawrence MW, Galindo Arranz M (2007) http://www.acoustics.org/press/135th/lawrence.htm

Lay T, Wallace TC (1995) Modern Global Seismology. Academic Press

Layton L (2007) Is it possible to test a nuclear weapon without producing radioactive fallout? http://science.howstuffworks.com/nuclear-test.htm

Le Pichon A, Blanc E, Hauchecorne A (eds.) (2009) Infrasound Monitoring for Atmospheric Studies. Springer

Massinon B (2004) Benefits of potential civil and scientific application of CTBT verification technologies. CTBTO Spectrum nr 4 July

Medalia J (2007) Nuclear Weapons: The reliable Replacement Warhead Program. CRS Report to Congress. Update February 8, 2007. Code RL 32929, Congressional Research Service

Medalia J (2008) Comprehensive Nuclear-Test-Ban Treaty: Issues and Arguments. Congressional Research Service Report for Congress. Order Code RL34394

Mikhailov VN (1996) USSR Nuclear Weapons Tests and Peaceful Nuclear Explosions. 1949 through 1990. The Ministry of the Russian Federation for Atomic Energy, the Ministry of Defence of the Russian Federation. RFNC-VNIIEF

MOD (2006) The Future of the United Kingdom's Nuclear Deterrent. Presented to the Parliament by the Secretary of State for Defence and the Secretary of State for Foreign and Commonwealth Affairs. December Cm 6994. http://www.mod.uk/DefenceInternet/AboutDefence/CorporatePublications/PolicyStrategyandPlanning/DefenceWhitePaper2006Cm6994.htm

NAS (2002) Technical issues related to the Comprehensive Nuclear Test Ban Treaty, Committee on Technical Issues Related to Ratification of the Comprehensive Nuclear Test Ban Treaty. National Academy of Sciences, National Academy Press, Washington DC, ISBN 0-309-08506-3 available at http://www.nap.edu

NASA (2007) Tectonic Plate Motion. http://cddis.nasa.gov/926/slrtecto.html

Nehru J (1954) http://www.indianembassy.org/policy/Disarmament/disarm2.htm

NEIC (2007) http://neic.usgs.gov/neis/station_book/

NEIC/USGS (2007) http://neis.usgs.gov/neis/gereral/magnitude_intensity.html

NOAA (2007a) http://www.pmel.noaa.gov/vents/acoustics.html

NOAA (2007b) http://oceanexplorer.noaa.gov/explorations/sound01/background/acoustics/media/sofar.html

Nordyke MD (2000) The Soviet Program for Peaceful Uses of Nuclear Explosions. September 1, Lawrence Livermore National Laboratory, USA. http://www.osti.gov/bridge/purl.cover.jsp?purl=/793554-ZAQEpq/native/

NRDC (2006) Natural Resources Defence Council, Press release. http://www.nrdc.org/media/pressreleases/061013.asp

NRDC (2007) Natural Resources Defence Council. Table of Indian and Pakistani Nuclear Tests 1975–2002. http://www.nrdc.org/nuclear/nudb/datab22.asp

NRDC Archive (2007) Natural Resources Defence Council. Archive of Nuclear data. http://www.nrdc.org/nuclear/nudb/datab15.asp

NSA (2006) The Vela Incident. Nuclear Test or Meteoroid? http://www.gwu.edu/~nsarchiv/NSAEBB/NSAEBB190/index.htm

Office of Technology Assessment US Congress (1988) Seismic Verification of Nuclear Testing Treaties. NTIS PB 88-214853

Office of the Director of Intelligence (2006) Statement by the Director of National Intelligence on the North Korea nuclear test. ODNI News Release 19-06 dated 16 October, Washington, D.C. http://www.odni.gov/announcements/20061016_release.pdf

Office of the Press Secretary of the White House (1995) Document 10149

O'Hanlon M (2008) Resurrecting the Test-Ban Treaty. Survival 50:119–132

OPBW (2007) http://www.opbw.org

OPCW (2007) http://www.opcw.org

Operation Shakti (1998) http://nuclearweaponarchive.org/India/IndiaShakti.html

Ramaker J, Mackby J, Marshall PD, Geil R (2003) The Final Test. A history of the Comprehensive Nuclear-Test-Ban Treaty Negotiations. CTBTO Preparatory Commission. Vienna

Richards PG (2004) Precision Seismology – Applications and Comments. http://adsabs.harvard.edu/abs/2004AGUFM.S11D.01R

Richards PG, Kim W-Y (2007) Seismic signature. Nature Physics 3:4–6

Richards PG, Zavales J (1990) Seismic Discrimination of Nuclear Explosions, Annual Review of Earth and Planetary Sciences 18:257–286

Richards PG, Zavales J (1996) Seismological methods for monitoring a CTBT. The technical issues arising in early negotiations. http://www.ldeo.columbia.edu/~richards/early CTBThistory.html

Richter CF (1958) Elementary Seismology. Freeman

Ringbom A, Elmgren K, Lindh K (2007) Analysis of radioxenon in ground level air sampled in the Republic of South Korea on October 11–14, 2006. Swedish Defence Research Agency report FOI-R-2273-SE

Ringdal F, Kværna T (1989) A multichannel processing approach to real time network detection, phase association and threshold monitoring. Bulletin of the Seismological Society of America 79:1927–1940

Ringdal F, Kværna T (1992) Continuous seismic threshold monitoring. Geophysical Journal International 111:505–514

Saey PRJ, Bean M, Becker A, Coyne J, d'Amours R, De Geer L-E, Hogue R, Stocki TJ, Ungar RK, Wotawa G (2007) A long distance measurement of radioxenon in Yellowknife, Canada, in late October 2006. Geophysical Research Letters 34, L20802, doi:10.1029/2007GL030611

Schaper A (2007) The fizzling fervency of the Comprehensive Test Ban Treaty. In Bremer Mærli M and Lodgaard, S (eds.) Nuclear Proliferation and International Security. Routledge

Schmidt (2007) Ultra-Fast Nuclear Detonation Pictures. http://www.waynesthisandthat.com/abombs.html

Schwitters R et al. (2003) Requirements for ASCI, October 2003, The Mitre Cooperation http://www.fas.org/main/content.jsp?formAction=297&contentId=256

Serber R (1992) The Los Alamos Primer, University of California Press

Shalikashvili JM (2001) Findings and Recommendations Concerning the Comprehensive Nuclear-Test-Ban Treaty. 4 January http://www.fas.org/nuke/control/ctbt/text/shalictbt.htm

Shearer PM (1999) Introduction to Seismology. Cambridge University Press

Shultz GP, Perry WJ, Kissinger HA, Nunn S (2007) A world free of Nuclear Weapons. Wall Street Journal 4 January

Sokolski HD (2004) Getting MAD: Nuclear Mutual Assured Destruction, Its Origin and Practice. http://www.strategicstudiesinstitute.army.mil/Pubs/display.cfm?pubID=585

Springer D, Denny M, Healy J, Mickey W (1968) The Sterling experiment. Decoupling of seismic waves by a shot-generated cavity. Journal of Geophysical Research 73:5995–6012

Stein S, Wysession M (2003) An Introduction to Seismology, Earthquakes, and Earth Structure. Blackwell Publishing

Sterngold J (2007) Los Alamos Scientist criticizes federal approach to arsenal. San Francisco Chronicle 13 February http://www.sfgate.com/cgi-bin/article.cgi?f=/c/a/2007/02/13/MNGI1O3N0G1.DTL

Strickland DA (1964) Report of the Conference of Experts to Study the Possibility of Detecting Violations of a Possible Agreement on Suspension of Nuclear Tests. Scientists as Negotiators: The 1958 Geneva Conference of Experts, Midwest Journal of Political Science 8(4):372–384

Stumpf W (1995) Birth and death of South African Nuclear Weapons Program. Atomic Energy Cooperation of South Africa Ltd http://www.fas.org/nuke/guide/rsa/nuke/stumpf.htm

Sublette C (2001) The Effects of Underground Explosions. http://nuclearweaponarchive.org/Library/Effects/UndergroundEffects.html

Sublette C (2007a) Nuclear Weapons Frequently Asked Questions. http://nuclearweaponarchive.org/Nwfaq/Nfaq10.html

Sublette C (2007b) Basic Principles of Staged Radiation Implosion. http://nuclearweaponarchive.org/Library/Teller.html

Subrahamanyam K (1999) A Strategic Overview. Embassy of India. http://www.indianembassy.org/press/New_Delhi_Press/August_1999/Strategic_Overview_August_09_1999.html

Surowiecki J (2005) The Wisdom of Crowds, Why the Many are Smarter Than the Few. Abacus

Sykes LR, Ekstrom G (1989) Comparison of Seismic and Hydrodynamic Yield Determinations for the Soviet Joint Verification Experiment of 1988. Proceedings of the National Academy of Sciences USA 86:3456–3460, May 1989 Geophysics http://www.pnas.org/cgi/reprint/86/10/3456

Top500 (2007) http://www.top500.org/

UNODA (1995) 1995 Review and Extension Conference of the Parties to the Treaty on the Non-Proliferation of Nuclear Weapons, 17 April – 12 May 1995, New York. http://disarmament2.un.org/wmd/npt/1995nptrevconf.html

US Department of State (2007) http://www.state.gov/t/ac/trt/5181.htm

USGS (2007a) http://www.earthquake.usgs.gov/research

USGS (2007b) http://www.earthquake.usgs.gov

Wall Street Journal (2007) The Wall Street Journal, February 22

Wallace TC (1998) The May 1998 India and Pakistan nuclear tests. Seismological Research Letters 69:386–393

Wisconsin Project (1996) Israel's Nuclear Weapons Capability: An Overview. http://www.wisconsinproject.org/countries/israel/nuke.html

Wotawa G, De Geer L-E, Denier P, Kalinowski M, Toivonen H, D'Amours R, Desiato F, Issartel J-P, Langer M, Seibert P, Frank A, Sloan C, Yamazawa H (2003) Atmospheric transport modelling in support of CTBT verification – overview and basic concepts. Atmospheric Environment 37:2529

Wüster J, Rivière F, Crusem R, Plantet J-L, Massinon B, Caristan Y (2000) GSETT-3: Evaluation of the detection and location capabilities of an experimental global seismic monitoring system. Bulletin of the Seismological Society of America 90:166–186

Zuccaro-Labellarte G, Fagerholm R (1996) Safeguards at research reactors, current practices, future directions. IAEA bulletin 38/4 http://f40.iaea.org/worldatom/Periodicals/Bulletin/Bull384/zuccaro.html

Name Index

Subject Index

Printing and Binding: Stürtz GmbH, Würzburg